Caryl Davis Haskins

Transformers

Their Theory, Construction and Amplification simplified

Caryl Davis Haskins

Transformers
Their Theory, Construction and Amplification simplified

ISBN/EAN: 9783337186203

Printed in Europe, USA, Canada, Australia, Japan

Cover: Foto ©Andreas Hilbeck / pixelio.de

More available books at **www.hansebooks.com**

ARTIFICIAL LIGHTNING. (See Appendix.)

TRANSFORMERS.

THEIR THEORY, CONSTRUCTION AND APPLICATION, SIMPLIFIED.

BY CARYL D. HASKINS.

ASSOCIATE MEMBER AMERICAN INSTITUTE OF ELECTRICAL ENGINEERS.

ILLUSTRATED.

1892.
BUBIER PUBLISHING COMPANY,
LYNN, MASS.

COPYRIGHTED BY
BUBIER PUBLISHING COMPANY,
LYNN, MASS.
1892.

PRESS OF
G. H. & W. A. NICHOLS,
LYNN, - MASS.

DEDICATION.

TO MY EARLIEST, MOST THOROUGH, AND MOST VALUED INSTUCTOR IN PRACTICAL ENGINEERING, MY FATHER, JOHN F. HASKINS, M. E., M. I. M. E., M. S. A., ETC., THIS LITTLE VOLUME IS RESPECTFULLY DEDICATED.

Caryl D. Haskins

CONTENTS.

DEDICATION.

	Page
CHAPTER I.	9

INDUCTION AND DISTRIBUTION BY ALTERNATING CURRENT.

CHAPTER II. 31

THEORETIC CONSIDERATIONS OF THE TRANSFORMER. — NON-REGULATION. — SELF-INDUCTION. — MUTUAL INDUCTION. — LOSSES, FOUCAULT CURRENTS. — HYSTERESIS, LEAKAGE.

CHAPTER III. 55

THE THEORY OF THE TRANSFORMER MATHEMATICALLY CONSIDERED.

CHAPTER IV. 63

EVOLUTION OF THE ELECTRICAL TRANSFORMER.

CHAPTER V. 73

TRANSFORMER CONSTRUCTION.

CHAPTER VI. 96

THE TRANSFORMER IN SERVICE.

CHAPTER VII. 110

COMMERCIAL TRANSFORMERS.

APPENDICES. 128

GLOSSARY. 143

INDEX. 145

PREFACE.

It has often been observed by almost every member of the electrical fraternity, that induction, and its outcome, the transformer, is to the popular mind, the greatest mystery of the whole lighting system with which they come in contact.

There is something tangible about the dynamo. Its movement, and the applied power are apparent. The lamp glows, and its action is appreciable, but the transformer remains to them an uncanny mystery.

So too the average electrician whose training has long accustomed him to the management and application of the electric current, finds in the transformer as a rule, more points regarding which his mind is hazy and uncertain, than in any other one piece of apparatus with which he has to deal.

The greater part of the printed matter dealing with the transformer which has from time to time appeared, has been either strictly technical, or entirely popular.

I have endeavored in the following pages to treat of the transformer and its action in such a manner as to render the work of especial value to the central station electrician, the student, and the investigator, while the greatest care has been exercised to render the matter so clear, simple and interesting, that it may come within the scope of the general public, and meet the demand for a semi-technical, and yet semi-popular treatise on the Electrical Transformer which has not heretofore been obtainable.

CARYL D. HASKINS.

Boston, August 1st, 1892.

TRANSFORMERS.

CHAPTER I.

INDUCTION AND DISTRIBUTION BY ALTERNATING CURRENT.

BEFORE taking up and considering in detail the calculation, construction and use, of the piece of apparatus which forms the subject matter of this treatise, it is necessary that the natural laws and commercial necessities which lead up to its manufacture and application should be briefly considered. It has been thought best therefore, to devote a preliminary chapter to the natural phenomenon know as "Electrical Induction," and to the general character, functions, and application of alternating currents. The object of this preliminary chapter, of necessity renders it of an elementary nature, but it can scarcely be considered superfluous, as it deals with the fundamental principle which underlies the whole subject. Since certain statements must in the consideration of all subjects be taken for established certainties, as bases for future argument, let us consider it an axiom that all electrical currents have a direction of flow, just as have all streams of water or currents of air. To render the idea of electricity more thoroughly consistent with our mode of thought, it will be found preferable to cease to consider it as an imperceptible existence, and to treat it as a tangible body; it will therefore be dealt with and regarded as a subtle fluid throughout the following pages.

It is a recognized fact that electricity as an imaginary fluid may under certain conditions be confined, and may exist in inertia; but at such times it is generally known as *difference of potential*, only when some form of motion results however, does electricity become manifest to human perception, and then only within certain ranges and under certain conditions.

The particular function of electricity with which we are chiefly to deal in this treatise is a very remarkable natural law known as Electrical Induction. Unlike many other electrical terms, the word induction is here very aptly applied. Induction, the noun, from the verb *induce*, which Webster defines as "to lead or influence by persuasion, to actuate, to impel, to urge;" "electrical induction" in fact, is that force which "persuades," "actuates" or "impels" an electrical current in one body by the influence of current in another.

To state this briefly it may be said, that *whenever an electrical current comes into existence* it induces or creates a current in *all surrounding* conductive masses, in a direction *opposite* to its own. During the period of its flow in uniform quantity, it has no inductive influence; the period of flow of the induced current extending *only* over a time equal to that between the commencement and attainment of full quantity of the inducing current. (This statement will be modified later, but for the present it may be safely accepted.) When a current *ceases* it induces a current in *surrounding masses*, in a direction *similar* to its own.

As has been already stated, currents are *induced* in "all surrounding conductive masses." This is strictly true, however remote these masses may be. The "making" or "breaking" of ever so insignificant a current circuit on the earth, undoubtedly induces currents in conductive masses on the *Moon*, but since the inductive influence *decreases* as the *square of the distance*, the effect, even through comparatively small intervals of space, becomes inappreciable. The interval of time, between the commencement of the current flow in the *inducing* circuit, or, as it is generally called, the "primary," and the commencement of flow in the *induced* circuit, or *secondary*, is known as the "lag" between primary and secondary.

Having briefly stated the general character of

inductive law, it will be well to consider the cause of this effect. This can be more clearly exemplified by resorting to simile. We do not claim the similes as original, but doubtless none more plain or conclusive could be found.

All active electrical conductors (for convenience we will limit our consideration of the matter to wires,) are considered to be surrounded by a field of force, little ripples or rings of *energy*, concentric with the conductor, extending equally in all directions, and becoming less and less intense as they become more and more remote from their source. By referring to diagram No. 1, this may be made more clear. Let A represent a section through the wire or conductor, and B, B', B'' the lines of force surrounding it; becoming less intense as they become more remote. Now the theory on which all inductive electric law is based is: That whenever a conductor *cuts through* "lines of force," or *vice versa*, when "lines of force" cut through a conductor, a current is set up in that conductor, provided always, that the conductor be part of a closed or complete circuit. Current effects may in reality be created in what would, generally speaking, not be considered a complete circuit. These effects are known as "Eddy" or "Foucault" currents, the former term exactly defining their character. Further consideration will be given to these effects in a later chapter.

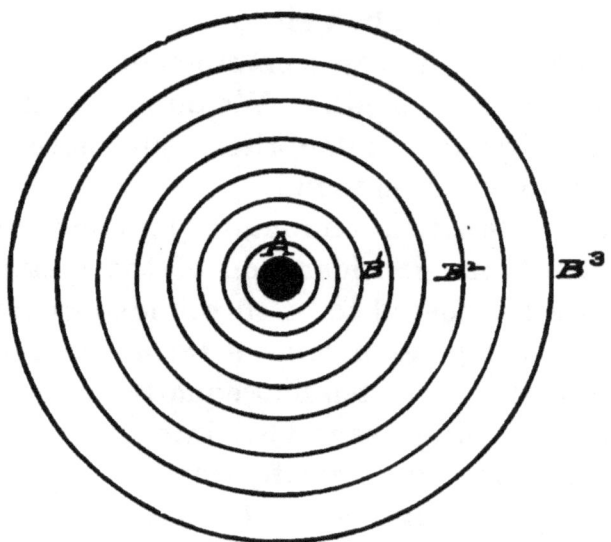

Fig. 1.

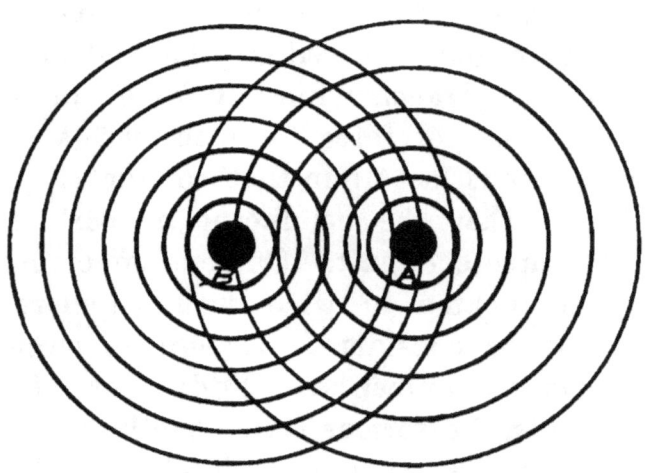

Fig. 2.

In the dynamo, current is generated by moving conductors, in such a way that they will cut through *lines of force*. In the transformer, current may be said to be generated by moving lines of force, in such a way that they will cut through conductors. Thus, in one sense, the dynamo and the transformer may be said to be opposites, although in reality almost identical in theory.

We will now endeavor to describe this whole matter more graphically. It should be borne in mind that in this, as in all similar descriptions throughout the volume, due allowance must be made for the liberty which is taken in treating of electricity, lines of force, etc., as actual palpable existences. In the case of the latter, at least, we may say quite frankly that they are simply influences which have been universally assumed to plausibly account for effects, the true cause of which we can only surmise. As has already been hinted, it is only the effects that are caused by electricity that are appreciable to us, the exact nature of the subtle influence itself is no more understood today than it was when the ancient Greeks discovered the effect of friction upon amber, and christened it *"Electron."*

Turning to Fig. 2, let us consider A and B to be electrical conductors, each one a portion of two separate complete circuits. A current springs into existence in A, the primary (no matter, for the present, what created it). Lines of force are immediately projected into the surrounding space, just as ripples are set in motion on the surface of water by a falling stone. These lines, in course of projection, must of necessity cut through B, the secondary. This *induces* current in B. Let us suppose that the current in A is flowing *towards* us; then the current in B is flowing *away* from us. As soon as the current in A becomes fixed in quantity, and the lines become *stationary*, the current in B *ceases*. Now let us suppose that the current in A ceases. The lines of force immediately collapse upon the centre B. In doing this they must necessarily cut through B again, but in an *opposite direction*. This sets up a second current in B, also in an *opposite direction* to the former one, so that the current in B is now coming towards us; the *same* direction that the current in A held until it ceased. Thus it will be seen, that, as has already been stated, when an electric current comes into existence it induces a current in neighboring conductors, in a *direction* opposite to its own. When a current ceases it induces a current in surrounding conductors, in a direction similar to its own.

Having made the fundamental principle of induction somewhat clear, we may now give some brief consideration to the system of electrical distribution, which this remarkable law has enabled man to formulate. Generally speaking, it may quite safely be said that the distribution of electricity through long distances is *not* economical, and for a very simple reason. The loss due to electrical resistance of the conductors must be considerable, since $C^2R =$ watts lost in transmission. Where $C =$ Current in Amperes, $R =$ Resistance of Conductor in Ohms. Thus it is evident that either the resistance of the circuit must be kept at a minimum, or the current strength must not be great.

The first of these two alternatives would at once necessitate the use of large masses of copper for conductors, but this is generally precluded by the high price of that valuable commodity. The other alternative, then, which in any case is the most desirable, is the only one worthy of consideration from a commercial standpoint.

To reduce the current strength or volume, and yet retain for transmission the same amount of force or energy, it of course becomes necessary to increase the voltage or pressure. In this respect electricity is quite analogous to water—the same amount of power may be delivered by a small stream at a high pressure, as by a large stream with small head. Thus far, then, the second alternative presents no difficulty, as by sufficiently increasing the pressure and reducing the current, high powers may be transmitted through long distances without the use of large quantities of copper and without excessive loss.

But a new difficulty now presents itself, inasmuch as currents of high pressure are extremely difficult to control and use for general purposes; besides which fact, it must be borne in mind that contact with high voltages is dangerous and sometimes fatal to human life. Incandescent lamps have not as yet been so constructed as to operate successfully on voltages in excess of 200 volts, which may be considered low pressure; whilst even if the lamps would bear higher voltage, such careful and perfect insulation would be necessary as to render their use commercially impracticable. It is plain, therefore, that the very pressure which renders it possible to carry electrical energy economically through long distances, would render it practically valueless, provided there were no way to reduce it at points of application without serious loss.

Induction provides us at once with a means of accomplishing this; as by proportioning the amount of wire in inducing and induced circuits, the energy delivered at a high pressure on the primary, may be transformed to a reduced pressure and greater volume in the secondary, or *vice versa*. This can of course only be accomplished by the use of intermittent, pulsating or alternating currents; any or all of which will produce the pulsations of the lines of force, which forms the fundamental principle of induction. The electric current which was first applied to incandescent lighting, was the direct, flowing constantly and uninterruptedly in one direction. This class of current does not of course permit of the use of induction for transformation, and has therefore to be distributed at useful or *low* pressures, which as we have seen renders it wasteful for long distance work.

As has just been stated, distribution by pulsating or alternating current, renders it possible to adjust the potential at any given point to any voltage, without serious loss. In practice however, only one of these classes of current is practicable, and this for reasons which will be presently stated, is the alternating.*

It may be well however, before going further, to make plain the difference between the pulsating and the alternating current. To state this quite plainly, it may be said that a pulsating current is one which starts into existence in direction "*a*" Fig. 3, dies down, more or less rapidly, then comes into existence again *in the same direction.*†
We have endeavored to make this a trifle more explicit, by the use of the diagram, the arrows

---

*This statement is made in a general way; there is at least one marked exception to the general rule. .

†"Pulsating," "intermittent," "vibratory," "interrupted," etc., are all names in common use for various modifications of the same class of current.

showing the direction of current flow during different periods.

In the technical consideration of alternating and intermittent currents, a system of "curve" drawing is used, whereby every rise and fall of current, every break and every alternation is accurately set forth. This system, which is almost invaluable

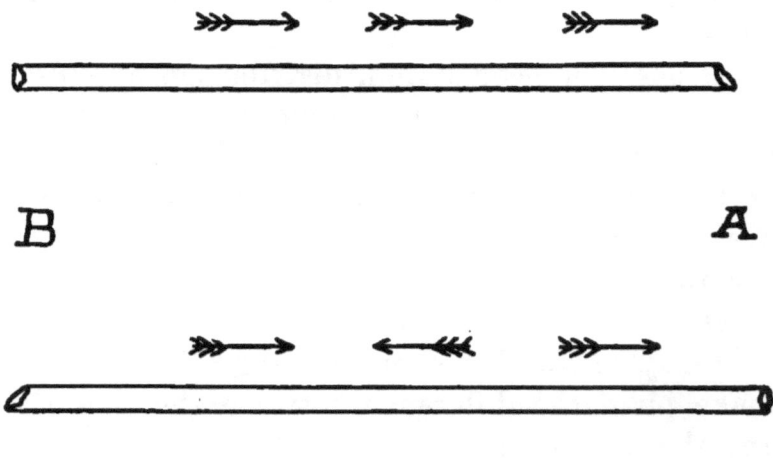

Fig. 3.

in all work connected with alternating or fluctuating currents, we will only briefly describe, as it is in general use, and almost all are familiar with it. Let us suppose that we have a rectangle, the surface of which is divided into very minute squares by vertical and horizontal lines. In drawing a curve showing the fluctuations of an alternating current, the distance between the vertical lines, or, as they are technically termed, "Ordinates," should be considered as fractions of time (generally taken in decimals of a second or minute), whilst the "Abscissae," or distance between the horizontal lines represent amperes (or fractions of current.) Two such curves are here represented. The "cross sectioning" is not shown, it being really unnecessary for purposes of description.

Fig. 4 represents an intermittent current. The line AB is the *neutral line*, and where the curve line touches this, no current is flowing. The portion of the curve above AB indicates that the current is flowing in, say $a\ +$ direction, whilst when the curve passes *below* the line, the current is indicated as flowing in a direction *opposite* to that which is held when the curve was above the line— or is flowing in, say $a\ -$ direction. AB is considered as 0, the amperes being numbered each way from it, those above being in one direction, those below in another.

It will be easy to imagine by casting the eye upon the upper curve, (the pulsating current) Fig. 4, how the lines of force will alternately be projected from and collapse upon the primary conductor, cutting surrounding conductors and inducing currents in them. The distance between the points where the curve touches the 0 line AB, shows us the length of duration of one period, or the speed with which the current is alternately

INDUCTION AND DISTRIBUTION. 21

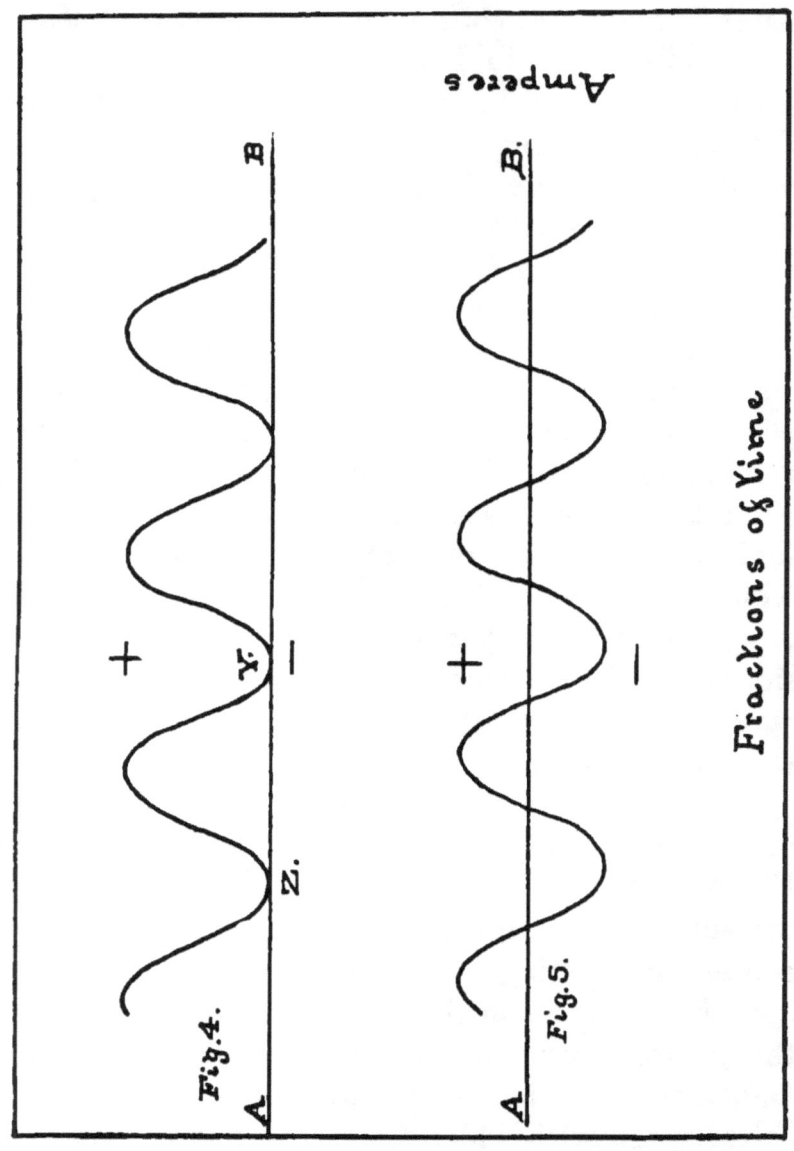

Fig. 4.
Fig. 5.

"made" and broken, whilst the distance between the 0 line and the maximum elevation of the curve, show us the maximum current strength. These curves are more often drawn to show the pressure waves, in which case of course, the abscissae indicate *volts*; or to show the *power* curve, the product of pressure and current ($C \times V =$ watts) in which case the abscissae become *watts*.

The same general discription applies equally well to the alternating curve shown in the lower position, Fig. 5. It is only necessary to bear in mind the fact that where the curve line crosses the 0 line, first a cessation and then a reversal of direction of current flow is indicated. It may be well to state that these lines follow very closely the rule of "sine" curves. We will make no attempt to discuss the theory of the sine curve as it has been thought best to confine the mathematics in this treatise exclusively to the principle and theory of the transformer.*

We may now return to our more immediate subject, and point out *why* the *alternating*, as compared with the pulsating current, is the only commercially practicable means of distributing large quantities of energy by the induction system. At first this seems rather incongruous, since we have already seen that the pulsating and the alter-

---

* For a concise, simple and clear elucidation of the "sine curve" we would refer the reader to Messrs. Slingo & Brooker's "Electrical Engineering," an excellent work.

nating current will "induce" equally well and efficiently, the former, in fact, probably more efficiently under certain conditions.

First and most important in considering this question, we must remember that the *natural outcome* of the dynamo-electric machine is an alternating current. It is the commutator, quite a distinct portion of the dynamo, which serves to "straighten" the current, and the commutator might quite as well be a distant machine, quite *separate* from the dynamo, so far as theory is concerned. It is only mechanical considerations and convenience which have caused it to be incorporated in that machine. The function of the commutator, in fact, is simply to reverse the connections at the *same instant* that the direction of current flow reverses, that is all. We may illustrate this: Suppose there were two tanks full of water, as illustrated in Fig. 6-1, A, B, A being higher than B, and that these two tanks were connected by a long rubber tube, easily removable from spickets, and that you held the end of the tube at X in the left hand, and the end Y in the right; the water would be running *from* A, through the tube in the direction of the arrow. Now suppose B were suddenly lifted *higher* than A, but at the same instant you pulled the end of the tube X off of A spicket, and thrust it on to B, and *vice versa* with the end Y, then the direction of flow would be

TRANSFORMERS.

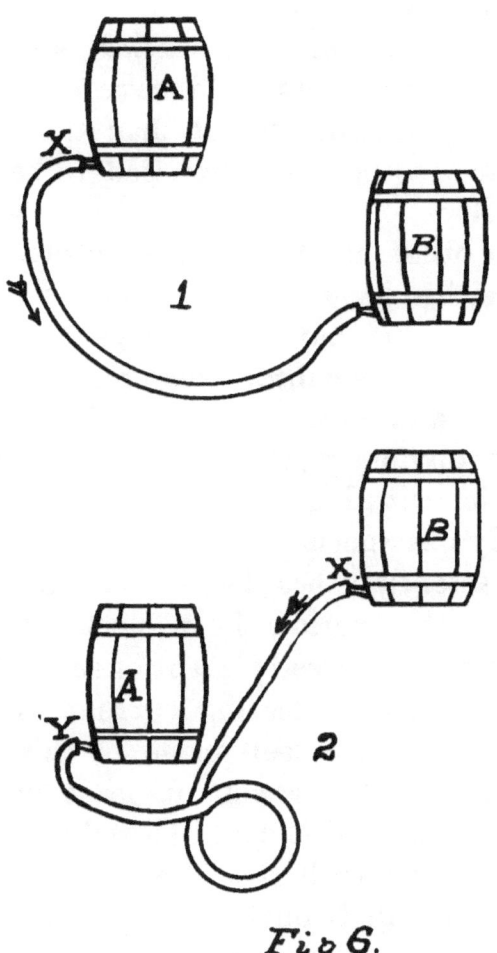

Fig 6.

maintained the same in the tube. It is just the function of your hands here that the commutator performs for the dynamo.

If the commutator was *omitted*, and collector rings substituted, we should have an alternating current output, *easier to get* than the direct, but not at first used because supposed to be less manageable.

On the other hand, to get a pulsating current we must either have a dynamo of special construction, or we must have an *interrupter* or circuit breaker which will vibrate very rapidly, alternately opening and closing the circuit. This latter device is obviously only practicable with *low* voltages, for with high potentials constant arcing, and speedy destruction, would *inevitably* result.

Much attention has been given of late by a number of the world's best electricians, to the special dynamos already mentioned as being designed to generate pulsating currents, and considerable advance has been made in this direction. The system would present at least one marked advantage, provided it could be reduced to a basis equally efficient with the alternating. It would permit of the distribution of power to electric motors, as well as for lighting purposes, a use to which alternating current has not as yet been applied with any marked degree of success, except in very small powers; in fact, the use of the alternating current is practically confined today to the distribution of light, and to a few special purposes; such as electric welding, etc.

It is easy to convert mechanical energy into electrical energy in the form of alternating current, but it is most difficult to convert alternating current energy *back again* into mechanical energy or *motion* with any degree of efficiency.

Pulsating or interrupted currents are largely used in connection with batteries and induction coils, where but small amounts of energy are to be dealt with, but not in connection with lighting work. Inductive coils and transformers, are practically the same thing, the former term is used in relation to coils for raising low potentials to high, especially where the powers dealt with are small, almost always in conjunction with batteries. Where *large* powers are raised to higher potentials, the coils are generally spoken of as Step-Up Transformers, which are treated of quite exhaustively in an appendix. The term Transformer, or as it is frequently called (especially in Europe) Converter, is used to specify coils intended for reducing high potentials to lower. Induction coils are extensively used in conjunction with batteries and an interrupter, for experimental work, medical purposes, electric gas lighting, the explosion of blasts, etc. They will be treated of in the following pages only in the most cursory manner, they are most ably described in a number of existing volumes, and scarcely come within our subject matter.*

Distribution by alternating current and transformer is commonly accomplished by the *multiple* or parallel system. Fig. 7 shows quite clearly the relative positions of dynamo and transformer when the latter are placed in multiple; with this arrangement the primary circuit, (and each secondary also,) has a constant potential, the amount of *current* varying with the load. Transformers have also at times been arranged in series, that is one after another, as for example in the old Jablochkoff system of arc lighting, but this arrangement is not common, and has only been used in connection with lighting service for street and arc lights. The series arrangement is clearly shown in Fig. 8. With the series system the potential is variable with the load, and the *current* constant.

The series and multiple arrangements of transformers may be clearly defined as follows: When transformers are placed in series the *same* current, and *whole* current, passes through the primary of each transformer one after another, and then back to the dynamo.

When transformers are arranged in multiple, a certain proportion of the entire current passes

---

* For information relating to induction coils, the author refers the reader to "Electricity and its Recent Applications," by Edward Trevert.

TRANSFORMERS.

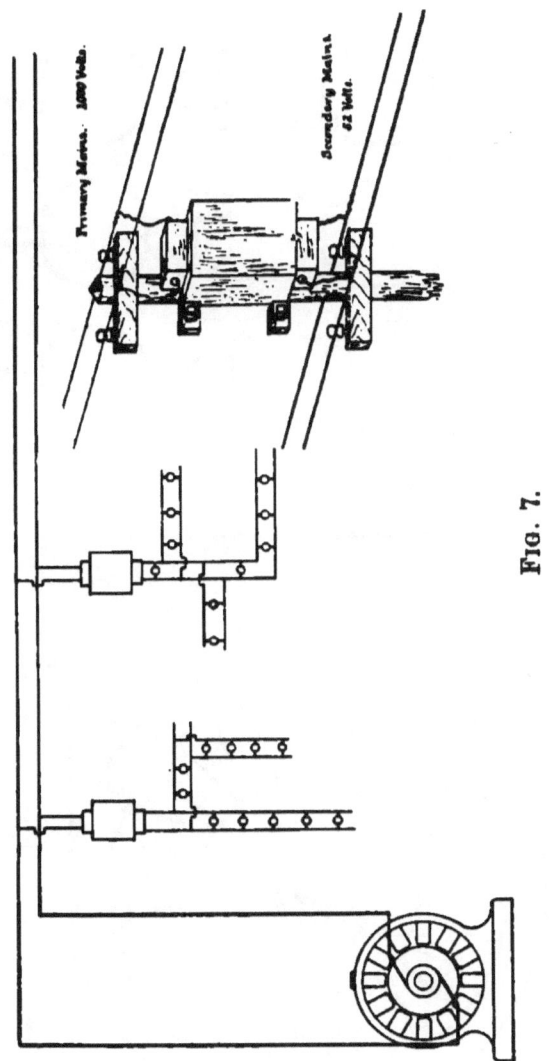

Fig. 7.

INDUCTION AND DISTRIBUTION.

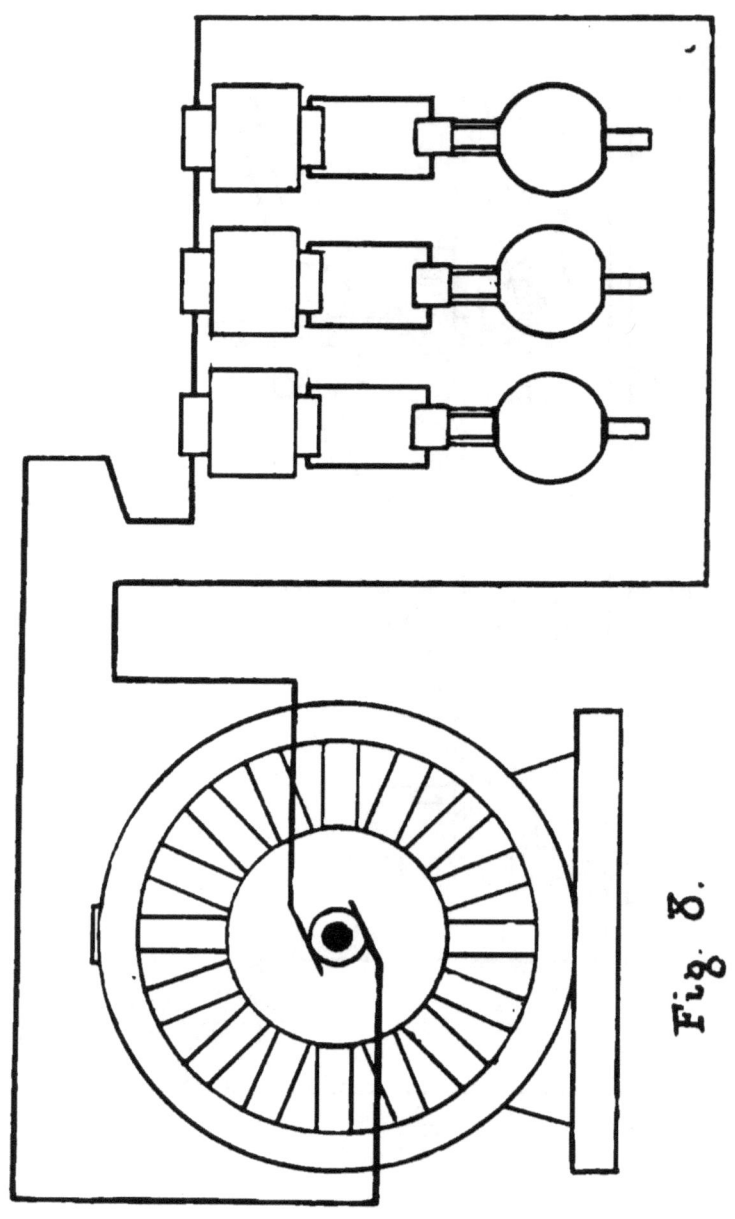

Fig. 8.

from the mains through each transformer primary individually, and thence back to the dynamo. If a single primary burn out or break with the series system, all of the lights go out, for the main circuit is open. (Special devices are however made to provide for this contingency.) If a primary burns out with the multiple arrangement only the lights on that particular transformer go out, or if the main circuit is broken anywhere along the line, then only those transformers which are *beyond* the break from the dynamo cease to operate. The lights which are between the dynamo and the break still continue to burn.

The multiple system is the only one in common use, and is the one with which we shall chiefly deal. The pressures, or voltages (primary), which are in common use are 1000, 2000 and 2500 volts, and in a few cases 5000 volts. Yet higher voltages than these are in use in Europe, but with these very high pressures proper insulation becomes a very difficult problem.

CHAPTER II.

THEORETIC CONSIDERATIONS OF THE TRANSFORMER.— IRON.— REGULATION. — SELF-INDUCTION. — MUTUAL INDUCTION. — LOSSES. — FOUCAULT CURRENTS.—HYSTERESIS.—LEAKAGE.

In the preceding chapter we have reviewed the principle and characteristics of induction, and the necessities and advantages which have led up to the use of the alternating current and the transformer, as an economic means for the distribution of light. We will now endeavor to treat somewhat in detail those various influences which serve to modify or enhance the phenomenon which we have already described.

As we have already seen, the creation or stoppage of a current in a conductor will induce a current in *surrounding* conductors, and it is known that this induction is subject to definite and known laws.

It is obvious that to bring long lengths of wire into one another's useful inductive influence, or "field" some other means must be adopted than that of stretching them side by side through a long distance, for this would be thoroughly impracticable, neither would this serve the purpose, even if convenient, for experiment has shown that the resistance of the two circuits being consistent, the voltage in the primary and secondary is almost exactly proportional to the respective lengths of the two circuits within one another's influence. Thus we may say, to state this arithmetically, that:

As influencing length primary : Influenced length secondary : : Voltage primary : Voltage secondary.

Now to bring, say ten feet of secondary, into the equal influence of one hundred feet of primary, both stretched in a straight line, is obviously impossible. It has been found necessary and most desirable in practice, therefore, to make the two wires into *coils*, placed one next to, or one over the other (see Fig. 9). In this way it is perfectly easy to bring a long wire into equal influence throughout its length, upon a short wire, AA being the short wire, the secondary (generally speaking,) and BB the primary, or long wire. But here is met a new contingency, that of *self-induction*. Let us consider this same coil (Fig. 9) in section Fig. 10. It is quite obvious that the lines of force projected from any turn of the primary B, must cut and influence, not only the neighboring turns of the secondary AA, but also the neighboring turns of *itself*, thereby setting up an influence in itself, in direct opposition to, and

THEORETIC CONSIDERATIONS. 33

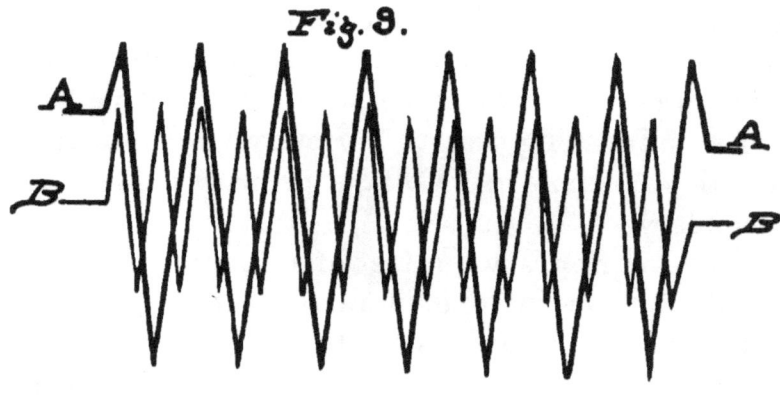

Fig. 9.

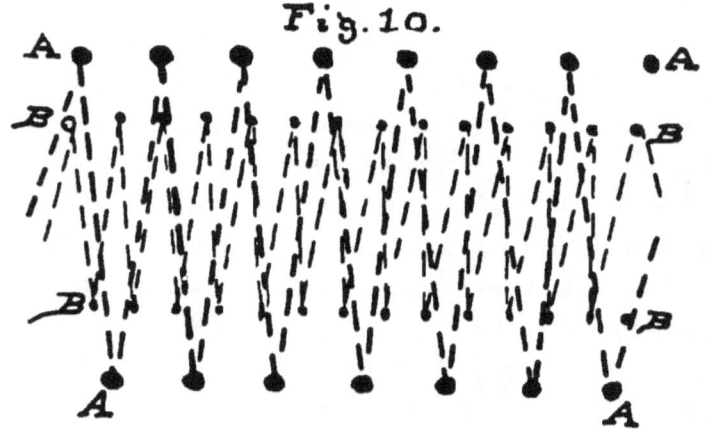

Fig. 10.

tending to *nullify*, the primary influence. Supposing for an instant that the secondary coil were absent, leaving the primary to act solely upon itself, turn upon turn, and supposing that nothing is lost in induction, or in other words, that the efficiency be 100 per cent, then the induced influence, the counter or opposing pressure (the "counter electro-motive force," as it is called), should exactly equal the *initial* primary force— one hundred units of force *pushing* against one hundred, and nothing would flow in the primary. A coil of this kind is used for some purposes, being known as an inductive resistance. At the first glance this would seem to entirely destroy the value of the transformer, but in reality it plays a most useful and important part when modified by the influence of *mutual* induction, as we shall presently see when considering the question of regulation.

Referring again to Fig. 10, we will now consider the part which the secondary coil plays. We have already seen how the primary induces current in the secondary, and opposing force by self induction within itself. It is obvious that since the primary coil induces current in the secondary, by reason of the lines of force projected from and retracted to itself, the secondary coil must in *its* turn react by induction upon the primary as soon as it becomes active, and this reactive induction of the secondary upon the primary is, of necessity, *favorable* to and in the same direction as the prime current of the primary coil and in direct opposition to, and tending to nullify the effect of the self-induction in the primary. Meantime however, the secondary has set up within itself, by its self-induction, a force in opposition to its prime or useful current, which is again practically nullified by the induction from the primary. Thus, the two coils act and react upon themselves and one another, the reaction of a coil upon *itself*, being termed as we have already said, *self-induction*, and that of one coil upon the other, which we have just discribed, *mutual* induction. The two terms are sufficiently indicative, being, in fact, almost descriptive of their relative action.

On these two factors of self and mutual induction, chiefly depends the successful operation of the transformer, for upon them alone rests the self-regulation of the transformer; without regulation it would be practically useless, because wasteful and inefficient.

Transformer regulation is simple and readily understood, and its presence amounts to a necessity.

In Fig. 11, we show a transformer which we will suppose to be in operation, P is the primary supplied from an active circuit, through which current is constantly flowing. S is the secondary,

36 TRANSFORMERS.

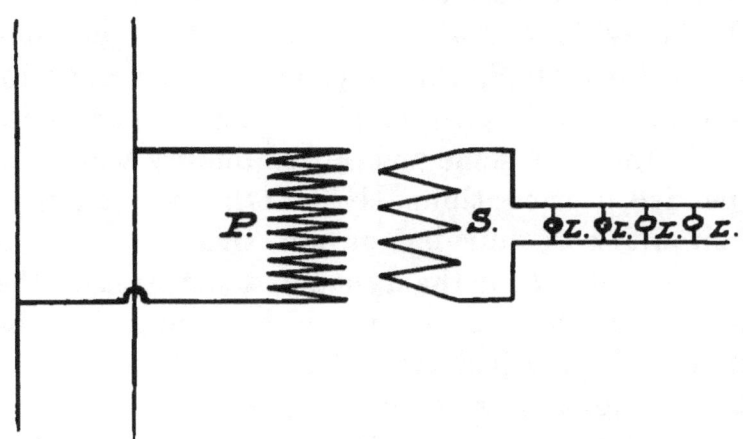

Fig. 11

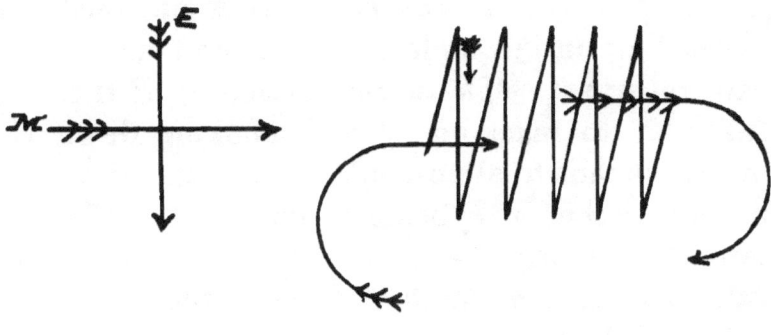

Fig. 12.

i. e. lamp circuit, and *L L* are lamps. The primary and secondary are separated from one another solely to render the diagram more distinct.

It is evident that if the amount of current flowing through *P*, were dependent solely upon its so-called "*dead resistance*," or the Ohms* in its circuit, then the same amount of energy would be expended in it continuously, whether the lamps in *S'* circuit were burning or not, this would be *extremely wasteful* for the dynamo, and therefore the engines at the station would be doing just as much work, and just as much *coal* would be burned when no lights were turned on, as when all were in use. Then too, so little current could be permitted to pass through *P*, that it could be practically of no use in transferring energy to *S*. Here self-induction comes into play. Supposing all of the lamps (*L*) to be turned off, *S* is obviously an "open" or an incomplete circuit, and no current can traverse it. The self-induction of *P* then has full play to react on *P* coil, choking down the prime current to almost nil, the amount of power, in fact present in *P* being about equal to the difference in energy between the prime current on full load and the reaction of self-induction.

This difference represents, practically, the *losses* in the transformer, with which we shall presently

---

* For a definition of terms which may be unfamiliar to the non-electrical reader, see glossary at end of volume.

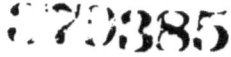

deal, almost all resulting, however, directly or indirectly in heat. The engine and dynamo are, therefore, doing very little work at this point, when no lights are in use.

Now suppose one light is turned on in S circuit. A little current now flows in S, dependent upon the resistance of S circuit. This sets up a slight mutual inductive reaction of S upon P, counterbalancing a portion of the self-induction of the primary. More current flows in P, and the engine and dynamo are doing just that much more work. And so the action goes on, as light by light is brought into service, till at full load (or the point of safe carrying capacity of the two wires) the self-induction of the primary is proportionally balanced by mutual induction. If the secondary coil were short circuited, as frequently happens, that is, if the two ends of S were joined without intervening resistance, such as lamps, then more current would pass than the wire could stand, and it would melt in two at its weakest point, quite as likely in the primary as the secondary. The same thing would happen if too many lamps were placed in the secondary.

To prevent the destruction of transformers in this way "fuses" are introduced, both in the primary and the secondary circuits. A fuse is a short piece of soft, easily fusible metal, usually lead and tin, calculated to melt, and thus break the circuit before the danger limit of the winding or wire is reached. When a fuse burns out, or "blows," all of the lights go out which are on that transformer, but nothing is injured.

The reader may have wondered why the curve of rise and fall of current shown in Fig. 5 of previous chapter did not abruptly break and commence, instead of gradually waxing and waning. This effect is due almost wholly to self-induction, and may be now readily understood. Self-induction may, in this connection, be looked upon as a kind of electrical inertia. A most ingenious device, known as a reactive coil, and used for turning lights up or down to any required brilliancy, is dependent for its action solely upon the effects of mutual and self-induction. It will be found fully described later, for it is an excellent example of these effects.

But we cannot deal with it until we have considered a matter on which we have not yet touched, but which is of prime importance: namely, the presence of iron in the transformer; on this entirely depends the efficient application of the phenomena already discribed.

We have dealt with *electrical* currents and their direction of flow: we must now introduce a new factor, that of *magnetic* currents, heretofore spoken of as lines of force. These are quite *distinct* from electric currents, but they go hand in hand, the one being dependent upon the other. Like electric currents they have (we suppose) a *direction,* we can scarcely say a direction of flow. Every known medium, even a vacuum is a conductor, but average *iron* is some 700 times better than any other known substance, all other mediums being about equal, save some comparatively rare metals of the iron group, such as nickel, cobalt, etc.

The direction of movement of these lines of magnetic force is at right-angles, to the direction of current flow, as shown at Fig. 12. With a single wire this "field" consists, as we have already seen, of concentric lines of force rotating around the conductor. These lines of force are of course purely imaginary, but something must be assumed to render the phenomena of electro-magnetism capable of being grasped. A very pretty demonstration of the presence of the force which we term magnetism may be made by thrusting a wire through a card, and then sprinkling the card freely with fine iron filings. On passing current through the wire, the filings will arrange themselves in concentric circles around the wire, exactly as we have described and shown in Chapter I.

Magnetic resistance is, in its way quite as disastrous to efficient operation, as is electrical; the *magnetic conductivity,* or as it is termed the *permeability* of the medium through which the lines of force are to be forced or circulated, has everything to do with the economical operation of all electro-magnetic apparatus. Thus in Fig. 12, No. 2, a single coil of live wire, the circulation of magnetic lines will be about as indicated by the arrows. Now if this coil be simple wire wound up hollow, or on wood, or brass, or in fact *any* substance other than metals of the iron group, then it will take, roughly speaking, some 700 times the electrical energy in the coil to maintain a given strength of magnetic circuit that it would if the magnetic circuit were of *iron* of sufficient sectional area. The intensity of the magnetic circuit is generally expressed by the number of "Kapp lines" per square inch of section, in other words the strength of the magnetic circuit, is supposed to increase according to the proximity of the lines to one another, that is, with the number within a given sectional area. It is obvious therefore that in constructing a transformer, the path for the magnetic circuit must be a complete circuit of *iron** if the apparatus is to be efficient.

As the resistance of the magnetic circuit increases with the length, this must of course be kept as short as possible, consistent with good mechanical construction. The permeability of iron also varies very greatly; this may be expressed by the number of Kapp lines per square inch, per

---

* This theory is contradicted provisionally by the inventor of the "Hedgehog" transformer, for which see appendix.

ampere turn* in a given length of magnetic circuit. Therefore it is necessary for the best results, to select iron of the greatest magnetic permeability, that is iron in a given sectional area of which the greatest number of Kapp lines can be induced with the least expenditure of energy.

The number of lines per square inch does not increase in a direct ratio with the number of ampere turns; there is a point at which the number of ampere turns must be vastly increased to secure even a slight increase of magnetic strength, and it is evident that it could be neither economical nor advantageous to build commercial apparatus to work, at, or above, this magnetic strength. The point at which the greatest number of lines per square inch can be induced with a relatively economical expenditure of current, is obviously the best magnetic strength to use in commercial apparatus. This is known as the "working point" of the iron, and varies greatly with the grade of material.

There is also another point, at which the magnetic strength practically ceases to increase, irrespective of any increase of ampere turns. This is the "point of saturation." At this point the number of Kapp lines per square inch of section have reached the limit of that particular piece of

---

* One complete turn of wire carrying one ampere, or two turns carrying ½ an ampere, etc.

iron. The point of saturation is never reached in commercial transformer practice.

In testing iron, all of the above characteristics are clearly expressed by plotting curves, similar to that shown in Fig. 13.

This curve shows wrought iron, having excellent qualities for transformer construction; its magnetic strength increases rapidly up to the working point as compared with the rate of increase of magnetic turns to obtain the result. It is of course understood by the reader that the greater the density of lines of force in the iron, the greater the energy induced in the secondary. Herein lies the whole question of transformer proportioning. How many turns in the primary, how many turns in secondary, and what section of iron to obtain certain results? This will be dealt with practically and theoretically further on.

Cast-iron gives the poorest results and is never used for transformers. Soft rolled charcoal iron has generally been considered the best, but the most recent practice seems to indicate that some grades of very soft rolled *steel* have the highest efficiency of all magnetic mediums—iron, however, is generally used.

A transformer constructed with cast-iron would be comparatively inefficient, or else the iron would have to be so increased in sectional area to obtain the necessary number of total Kapp lines, that

TRANSFORMERS.

Fig. 13

bulk and weight would be prohibitively large. In other words, the working point of this iron is far too low.

Before considering transformer losses, mention should be made of the reactive coil already referred to, for in it is embodied a practical illustration of the application of all of the laws as yet referred to.

Fig. 14 shows one of these coils, which consists of three main portions, a ring of iron highly laminated (A), around which is wound a considerable number of turns of wire, in series with the primary or secondary circuit which it is to control. A circular block or plug of iron (D), also highly laminated, fitting closely, but without touching, into the hole through A, and supported and rotating in bearings at F. A solid copper casting (C), forming a complete secondary circuit of one turn, rigidly fastened to D. C is provided with a handle. D and C are, of course, movable in the bearings F, and can be placed in any position in the arc of A. Now when C is turned until directly opposite B, it is obvious that the lines of force set up by the current in B will not pass around A, but will take a short cut across through D, and thus round and round, for lines of magnetic force, like electricity, always take the path of least resistance. Since the lines of force do this, C, in its present position does not come within their influence, and remains inactive. Thus the current in B reacts solely upon *itself*, and by self-induction, as already explained, chokes down the flow of current, and the lamps in its circuit glow very dimly. As C is drawn over towards B, it begins to enclose more of the lines of force, whereupon current begins to flow in it, and it becomes a secondary. The secondary current in C, reacts upon B, by " mutual induction," and permits more current to flow, the lamps becoming brighter. The nearer to B, C is drawn, the more lines it encloses, the more current is set up in it, and the more it reacts on B, till when C rests directly over B, all of the lines of force are brought into play, the action and reaction of mutual, and self-induction about balance, and practically full current flows through B to the lamps.

We have now to consider the various causes of *loss*, incident to the transfer of energy, from the primary to the secondary of a transformer. First, and most serious is the C^2R loss in the primary and secondary. This is dependent solely upon the Ohms resistance of the two coils, and is in accordance with Ohms law with which we assume the reader to be familiar. This loss is unavoidable, but may be kept at a minimum by using as few turns as possible to accomplish a given result, and by using copper of great purity, and of ample sectional area to carry the required current. The

THEORETIC CONSIDERATIONS.

FIG. 14.

C^2R loss may be computed by means of Ohms law, after measuring the Resistance (in Ohms) of primary and secondary coils.

The next important loss in the transformer is due to Eddy or Foucault currents, which occur in the iron, and to some extent in the coils themselves. As has been stated in the foregoing chapter, the character of these currents is pretty clearly emphasized by their name. They are very similar in character to the local eddies and currents which are to be found everywhere in running streams of water. The iron of transformer cores being of course conductive, current in the primary coil has an inductive effect upon it, quite apart and distinct from its magnetising influence; this effect is quite analagous in character to the induction of current in the secondary, small local currents of electricity being set up throughout the mass of iron. Energy is naturally required to generate these Eddy currents, and this energy is taken from the primary and secondary circuits, which induce them, whilst as these Foucault currents do no useful work, the force expended in generating them is absolutely wasted. If steps were not taken to prevent the presence of Eddy currents in the core, in dangerous quantities, they would at once prove disastrous to the efficient operation of the transformer; this would be the case if the iron were one solid mass, for they would then have a free path through which to circulate. Fortunately it is a comparatively easy matter to guard against and prevent the generation of Eddy currents in dangerous quantities by simply so subdividing the core that the path through which these eddys would natually flow, is broken at short intervals by minute non-conductive spaces, or spaces filled with matter of very high electrical

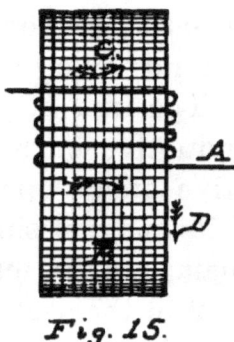

Fig. 15.

resistance. Since the potential of Foucault currents is always very low, these insulating spaces may be comparatively imperfect.

Since Foucault currents must always have a direction consistently in the plane of direction of the inducing current; and since the magnetic circuit follows a direction at *right-angles* to the electric circuit, it is only necessary to build up the iron core of thin sheets, or of a large number of turns of iron wire partially insulated from one another by tissue paper, corrosion, or other means, to effectually prevent the presence of Eddy currents in dangerous quantities, without breaking or interrupting the path of the lines of force. This may be better understood by referring to Fig. 15, A being a coil of wire carrying an alternating current, and wound around an iron ring, B. The arrows, C, indicate the plane of direction of Foucault currents, whilst the arrows, D, show the plane of direction of the magnetic circuit. The iron ring is made up of a number of punched washer-like pieces of sheet iron, separated by tissue paper or other means, and closely clamped together. It will be seen that the path of the Eddy currents is obstructed at short intervals by the layers of insulation, whilst the path of the magnetic lines is closed and unobstructed. The same would hold good if the core was constructed of a bundle or coil of iron wire, and such a core was quite largely used in early practice, but a laminated core of sheet iron is now almost universal, because, aside from economy, the magnetic discontinuity of a closed circuit wire core is obvious, besides which, plates or punchings are much more readily clamped firmly together.

Practically all of the energy which is present in the core in the shape of Foucault currents, is expended in heating the iron; this heat is objectionable, since it warms the copper wire of primary and secondary, and since the resistance of copper increases with the rise of temperature, the C^2R loss is magnified, and it becomes necessary to use somewhat larger wire than would be required if there were no heating.

Minute Eddy currents are also induced within the copper of primary and secondary, being quite distinct from the prime currents, but, except in transformers, having secondaries of large current capacity and containing therefore, considerable masses of solid copper, their influence is almost negligable, tending, however, to increase the temperature of the coils. There are several excellent formulae for computing the losses due to Foucault currents. Being somewhat complicated and involved, however, they have been omitted.*

Foucault current losses, in combination with other heating effects, may be closely and easily computed by means of the calorimeter, which will receive further mention a few pages later. It is obvious that, in view of the prejudicial effect of heat, due allowance should always be made for it, and such opportunity for radiation be provided as is consistent with mechanical construction.

Hysteresis, the last of the loss producing forces to be considered, is far less clearly defined in character, and much more difficult to grasp, than any of the effects as yet treated of. It is only

---

* For an able theoretic treatise on Eddy currents, we would refer the reader to the Phil. Mag., (England) Jan. and Feb. 1884.

very recently that hysteresis has been individualized and separated from other influences of similar effect.

Hysteresis may probably justly be considered as being the direct outcome of magnetic inertia. Loss by hysteresis may, in fact, be looked upon as the energy used in overcoming a kind of internal friction between the molecules of iron, which, according to the accepted theory, change their position every time the polarity of the iron is reversed, and this theory may the more readily be accepted, since the loss due to hysteresis decreases with the *increase* of mechanical vibration, which serves to agitate, and therefore assists in moving the molecules. Let us suppose a piece of iron to be so enormously magnified that the eye can distinguish the molecules, one from another, the whole mass having the appearance of an enormous number of particles grouped together and held in the form of a homogeneous mass only by molecular attraction, as is really the case. When this mass is subjected to the influence of an electric current, it becomes magnetically polarised, that is, *each molecule* has a north and a south pole. Now if the current be reversed in direction, the *magnetism* necessarily reverses, *not*, as it might seem, by the molecules being first demagnetized and then magnetized in an opposite direction, but by turning all of the molecules half-way round, so that their existing poles may coincide with the reversed conditions.* This reversal is necessarily accomplished by the expenditure of a certain amount of energy, which results directly in *heat*, due (we assume) to *molecular friction*. This waste of energy, is the loss due to hysteresis. Hysteresis losses naturally increase with the frequency of the alternations, and almost in a direct ratio.

The heat losses due to Foucault currents, and to hysteresis, can most conveniently be ascertained by the Calorimeter test. The principle of this method is simply the measurement of the rise in temperature due to these effects, whose value is then established by comparison with a scale of temperatures due to known expenditures of energy. The transformer, or other piece of apparatus in which these losses are to be measured, is placed within a closed box, whose temperature is normally fixed, owing to its being packed in ice, or surrounded by circulating water or air. The total rise in temperature above normal, of the air within the box, then represents the total heat losses.

This method, which is extremely exact, is not more fully described, as it is commonly used in the most careful and exact tests, as carried out in the leading technical schools.

---

* It must be borne in mind that we are treating of a theory, which is however, generally accepted.

The commoner method (though less exact) is to measure the *total* losses, by simply ascertaining the watts expended in the primary, when the secondary is idle, due allowance being made for magnetism and C^2R losses. The per cent of loss (at full load) in the average transformer of today varies from about 10 to 15 per cent in very small, to 3 to 5 per cent in large converters. It is probably easier to design a transformer of, say 10,000 watts capacity, with an efficiency of about 97.5 per cent, than a 250 watt converter, with an efficiency of 90 per cent. For this reason the fewer the number and the greater the size of the transformers, for a given number of lights, the greater the efficiency of the system, provided, always, that the converters are working at full load in both cases. A converter is, of course, most efficient at full load, for the waste (except the C^2R losses) is practically a fixed quantity at all loads. For this reason transformers should not be installed, having a capacity in *excess* of what they are called upon to supply. Further information relative to this will be found in chapter VI.

The reader may note that the question of lag has been neglected. This has been done intentionally, and after due consideration, as purely technical matters do not seem to be in keeping with the character of the work in hand.

CHAPTER III.

THE THEORY OF THE TRANSFORMER MATHEMATICALLY CONSIDERED.

WHILST it has been deemed advisable to avoid, as far as possible, all purely technical matters in this work, it could scarcely be considered complete without at least a brief consideration of the mathematical theory of the converter. After due consideration Hopkinson's Formulæ for the Transformer, as set forth by Prof. S. P. Thompson, has been selected, as being at once the simplest and the most direct.

The following equations are not, in the remotest sense, original in any respect. They are presented purely for the convenience of the reader, and in the simplest form possible. A knowledge of algebra only is presumed.

The following symbols have been assumed:

A—Sectional area of core in square centimeters.
E_1—The whole applied electro-motive force in primary.
C—Current in amperes in external circuit.
C_1—Current (absolute C. G. S. units) in primary.

C_2—Current (absolute C. G. S. units) in secondary.
H—Intensity of magnetic field.
L_1—Coefficient of self-induction in primary coil.
L_2—Coefficient of self-induction in secondary coil.
l—Length of magnetic circuit in core.
M—Number of magnetic lines per square centimeter. The magnetic induction.
N—Total number of lines of force traversing the core.
p—Coefficient of transformation—ratio between windings.
R—Resistance of external circuit in ohms.
R_1—Resistance of primary in ohms.
R_2—Resistance of secondary in ohms.
r_2—Internal resistance of secondary in ohms.
S_1—Number of turns of wire in primary coil.
S_2—Number of turns of wire in secondary coil.
t—Time measured in seconds.
μ—Coefficient of magnetic permeability of iron.

It is, of course, understood by the reader that π is the accepted symbol for the ratio of the circumference to the diameter of a circle, which equals about $22/7$ of the diameter; or, more exactly, 3.14195.* Hopkinson's theory of the transformer has, as its basis, the magneto-motive forces; or, to state it more clearly, the applied magnetism-creating energy at work in the iron core. Considered as

---

* It is well to remember the following expressions: Circumference of a circle = $2\pi r$. Area of a circle = πr^2. r, denoting the radius.

resulting from the algebraic sum of the ampere turns in the primary and secondary, from this the voltages resulting from the variations in the magnetic induction of the core, are deducted.

Hopkinson first considers the magneto-motive force necessary to project N lines of force through a core, having 1 length and μ permeability—this, of course, being entirely dependent on the quality of the iron. The total number of magnetic lines

$$= \frac{\text{magneto-motive forces.}}{\text{magnetic resistance.}}$$

or we may, according to Hopkinson's formulæ, write:

$$\tfrac{4\pi}{10}(S_1 i_1 + S_2 i_2) \div \tfrac{l}{A\mu} = N. \quad (1)$$

If we know the cross sectional area of the iron (A), and the number of lines per square centimeter (B), we may say:

$$N = AB \quad (2)$$

in which case the magneto-motive force is

$$\tfrac{4\pi}{10}(S_1 i_1 + S_2 i_2) = AB\tfrac{l}{A\mu} = \tfrac{B}{\mu}l = Hl \quad (3)$$

Next state two equations for the pressures created in the primary and secondary circuits, as follows:

$$\begin{cases} E_1 = R_1 i_1 - S_1 \tfrac{dN}{dt} = R_1 i_1 - S_1 A \tfrac{dB}{dt} \quad (4) \\ 0 = (r_2 + R_2)i_2 - S_2 \tfrac{dN}{dt} = (r_2 + R_2)i - S_2 A \tfrac{dB}{dt} \quad (5) \end{cases}$$

R_2 being the resistance of the lamp circuit and r_2 the resistance of the secondary coil itself. We now multiply the fourth equation by S_2 and the fifth by S_1, getting:

$$S_2 E_1 = S_2 R_1 i_1 - S_1(r_2 + R_2) i_2 \quad (6)$$

this, with the third equation, gives us:

$$(7)$$

$$i_1 \{S_2^2 R_1 + S_1^2(r_2 + R_2)\} = S_2^2 E_1 + S_1(r_2 + R_2)(10 Hl \div 4\pi)$$

$$i_2 \{S_2^2 R_1 + S_1^2(r_2 + R_2)\} = -S_1 S_2 E_1 + S_2 R_1 (10 Hl \div 4\pi)$$

and

$$A \frac{dB}{dt} = -\frac{(r_2 + R_2) S_1 E_1}{S_2^2 R_1 + S_1^2 (r_2 + R_2)} + \frac{10 Hl}{4\pi} \cdot \frac{R_1 (r_2 + R_2)}{S_2^2 R_1 + S_1^2 (r_2 + R_2)} \quad (8)$$

The second term $(_8)$ may be ignored, since Hl is insignificant in comparison with $\frac{4\pi}{10} S_1 i_1$; it being only the difference between $S_1 i_1$ and $S_2 i_2$ (which are large factors), that is needed as a magneto-motive force, which is small, provided the permeability of the iron is great as is customary in actual practice.*

We have, therefore:

$$A \frac{dB}{dt} = \frac{dN}{dt} = -\frac{(r_2 + R_2) S_1 E_1}{S_2^2 R_1 + S_1^2 (r_2 + R_2)} \quad (9)$$

bearing in mind that:

$$E_2 = S_2 \frac{dN}{dt} \quad (10)$$

---

*It is not generally customary to carry the number of lines much above 10,000 lines per square centimetre.

we have:

$$E_2 = -\frac{(r_2+R_2)E_1\frac{S_2}{S_1}}{\frac{S_2^2}{S_1^2}R_1 + r_2 + R_2} \quad (11)$$

Since the effective electro-motive force in the secondary circuit (E_2) is equal to $(r_2+R_2)i_2$, it follows that:

$$i_2 = -\frac{E_1\frac{S_2}{S_1}}{\frac{S_2^2}{S_1^2}R_1 + r_2 + R_2} \quad (12)$$

Assuming (as is practically true) that the voltage at the terminals of the primary coil of the transformer is constant, we may take E and R, as relating purely to that which is internal in the primary coil; and it is quite plain that the action of the converter is to reduce (or increase) the pressure in direct proportion to the ratio of the winding, for example, to reduce a primary pressure of 1,000 volts to a secondary E. M. F. of 50 volts we might say, 50 : 1,000 :: 100 : x or 2,000 turns of primary to 100 turns secondary, or 20 to 1, following a like argument in all required secondary voltages. We also see from the foregoing arguments that the effect of the transformer is to add to the resistance of the secondary circuit a term equal to the resistance of the primary, *reduced* in proportion to the *square* of the ratio of the windings. It is also obvious that, owing to the negative sign in equation number twelve, the two currents are in exactly opposite phases.

It is a general, and a pretty safe, rule to assume that the weight of copper in primary and in secondary are equal, this being so the following table of relations and proportions will be found to be correct. We quote this table from "Dynamo Electric Machinery" by Prof. S. P. Thompson.

	PRIMARY.	SECONDARY.	RATIO.
Windings.....	S_1	S_2	p
Resistance....	r_1	r_2	p^2
Self-Induction...	L_1	L_2	p^2
E. M. F......	E_1	E_2	p^2
Current.....	i_1	i_2	p
Heat Waste....	$i_1{}^2 r_1$	$i_2{}^2 r_2$	1

As has been seen, the question of proportioning the iron, depends almost absolutely on its permeability, and the length of the magnetic circuit, the following weight, however, is a fair average for the various medium sizes of standard commercial transformers.* Weight, per useful applied watt at secondary terminals equals .103 lbs. This includes weight of copper, etc., which must be deducted to get iron weight where required. Very small transformers are heavier in proportion, whilst very large ones are generally much lighter.

---
* 15 to 30 lights.

MATHEMATICAL CONSIDERATIONS. 61

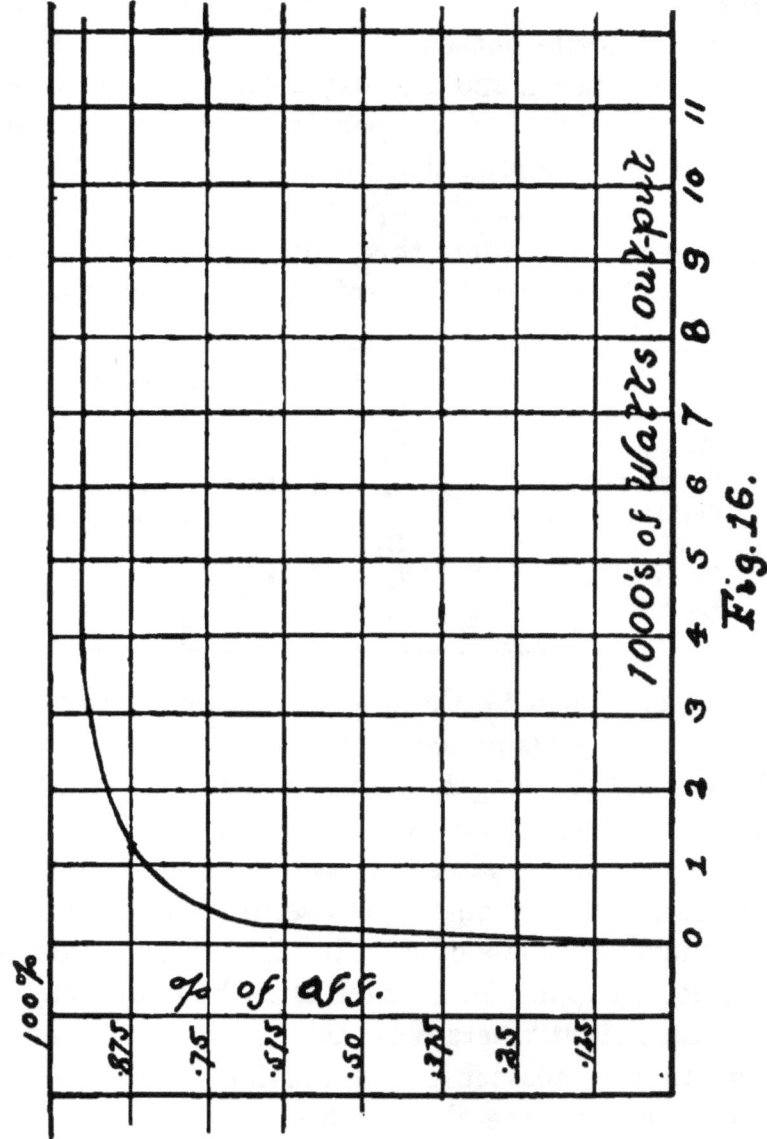

Fig. 16.

It is of course, understood that the efficiency of a transformer varies with the load, the least efficient point being the lowest load, and the most efficient, the highest. The efficiency is graphically shown by plotting the results of tests as an efficiency curve. A good example of this is shown at Fig. 16.

The horizontal lines representing the per cent of efficiency, and the vertical lines, the load in thousands of watts. In this particular case it will be noted that the maximum efficiency shown is 97 per cent, the remaining 3 per cent representing the losses. The potential of the secondary also varies somewhat with the load, it being highest on very low loads, and lowest on full loads. The average variation between one lamp and full load, being about 1.9 volts, in average sizes of converter.

CHAPTER IV.

EVOLUTION OF THE ELECTRICAL TRANSFORMER.

In these days the alternating system has come into universal use. The high potential current of the dynamo is carried great distances with but little loss, and is reduced to the low pressure necessary for house lighting, by means of a transformer—one of a dozen or more kinds—all of them efficient, sure in their action, and entirely reliable. Transformers are sold as a mere commercial article, and lighting companies order a dozen of them as a cook might a dozen eggs.

In view of all these facts, it is, perhaps but natural in us to forget that the success of the transformer, in fact the discovery of its principles, is a matter of comparatively recent date, and that it has gone through very many changes before reaching its present comparatively perfect form. Many people, if asked who first invented and used the induction coil, would answer, if they were only casual readers on the subject—"Ruhmkorff." It is very easy for us in these days to forget Faraday and his discoveries. Still easier to forget that the origination of the transformer is due to him. Yet the original transformer of Faraday embraced all of the essential features of the best transformers of today.

Fig. 17, represents the original "Perfected Induction Coil" of Faraday. A is a ring of cast-iron,

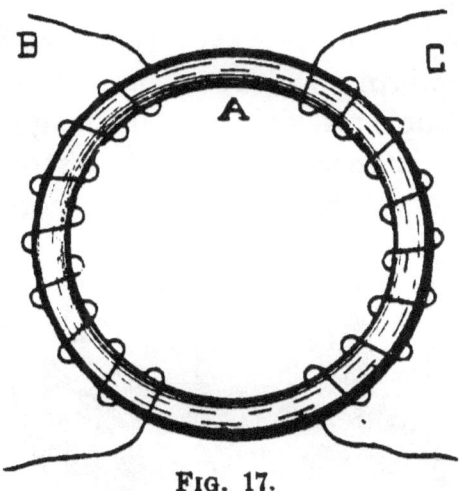

FIG. 17.

B the secondary, C the primary. It will readily be seen that we have here the most essential feature of the modern transformer, namely,—a closed magnetic circuit.

The Ruhmkorff induction coil was invented and patented in 1842 (see Fig. 18).

Its construction is familiar to all—a straight rod of iron forming a core, a primary wound upon it, and a secondary upon that.

This coil was designed, and is very extensively used for inducing high from low potential intermittent or interrupted currents, in connection with batteries, in laboratory work, and also for medical and electro-chemical purposes.

The Faraday and Ruhmkorff coils, are types of the two classes of converter prevailing today—the opened and the closed circuit transformers. It will readily be seen however that the Ruhmkorff, having a straight core, must complete its energetic circuit through the air; thus its magnetic circuit is of high resistance and presents a strong contrast to the Faraday coil, which has a complete mag-

FIG. 18.

netic circuit of iron, and hence of low resistance. Generally speaking, the lower the resistance of the magnetic circuit, the greater the efficiency of the transformer, provided other parts are properly proportioned. The converters of the Faraday type are therefore the more efficient, and the transformer of today is almost universally of this class.

To shorten and lower the resistance of the magnetic circuit is one of the chief aims of the modern transformer builder. This is done in four ways: first, by making the circuits entirely of iron; second, by shortening these magnetic circuits as much as possible; third, by increasing the sectional area as much as is consistent with weight; and fourth, by using iron of the greatest magnetic permeability. With these features always in view, and due care with regard to insulation, and the best ventilation possible, a transformer with properly proportioned windings, can be made extremely efficient. It is interesting to note the various changes and improvements which the transformer has passed through in attaining its present high character.

Jablochkoff and Goulard & Gibbs, used transformers of the Ruhmkorff type in the first distribution which was attempted by the alternating system. These early transformers were used in series, a practice which has long since been discontinued except in connection with one or two systems of street lighting, and for special purposes. Probably the first improvement on the old Faraday coil (and it was a very slight one), was made by Kennedy in '83; he employed the same large iron ring, with alternate windings of primary and secondary, the improvement being in the ring, which was of iron wire instead of cast-iron, thus gaining greater magnetic permeability, it being wrought iron highly laminated. Ferranti somewhat improved this a little later, by winding strip iron around the periphery over the coils. This somewhat reduced the magnetic resistance, since it increased the cross-section of iron employed; it also served to inclose any magnetic leakage which might occur. The merits of this device were questionable however, the additional iron necessarily increased the weight, whilst the ventilation

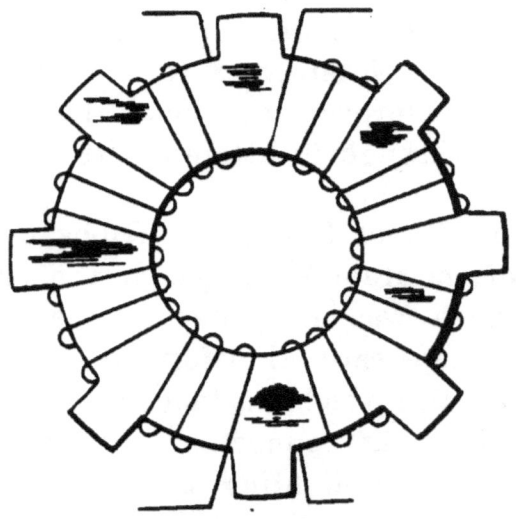

FIG. 18 A.

being reduced, heating naturally resulted, and the liability of a "burn out" increased.

Stanley a little later made a transformer, which he claimed to be a marked improvement over anything then existing. He made a ring of finely laminated iron of the shape shown in Fig. 18 A.

This was an adoption of the Paccinotti armature. It is difficult to say what were the advantages expected to result from this construction. The leakage from the teeth of the ring was anything but desirable, a considerable amount of magnetic energy doubtless being thus dissipated. It however served to give an idea to Dick and Kennedy,

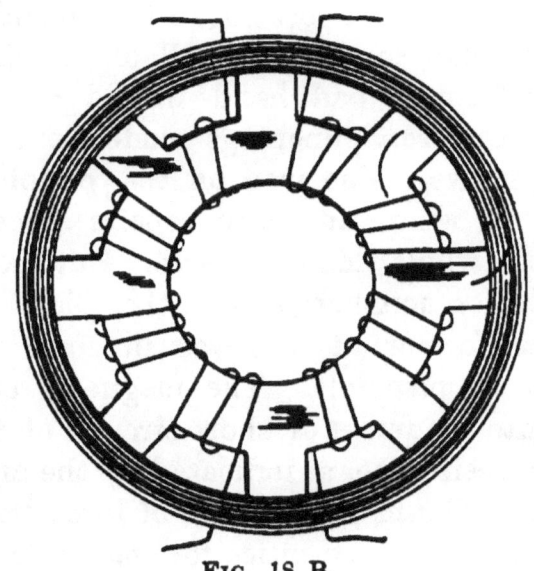

Fig. 18 B.

who in '86 introduced a transformer which showed the first really vital improvements since the days of Faraday.

Starting with Stanley's Paccinotti ring, they wound the periphery with thin sheet iron, the result being a transformer in appearance like Fig. 18 B. A coil of considerable efficiency and correct principle. Much of the efficiency which should have been gained by the reduction of the length of the magnetic circuits, was sacrificed by the magnetic resistance due to these circuits not following the direction of the lamination in the peripheral iron.

But at this point in the history of the transformer, the question of cost began to demand serious attention. Up to this time all of the closed circuit transformers had hand wound coils. The shape of the iron punchings made them wasteful and expensive. The iron on the periphery was difficult to wind, and any repairs necessitated tearing the whole thing to pieces. But though too expensive for commercial use, the Dick and Kennedy transformer of '86, was in some respects excellent in principle. The magnetic core was cut up into a number of short circuits of theoretically low resistance, as indicated by the arrows in Fig. 18 B. The magnetic lines of force were quite well enclosed, opportunities for magnetic leakage being almost entirely absent, and the ventilation was good. A number of tests of this transformer carefully conducted, proved its efficiency to be fairly high. There is a further objection to this coil however, which applies equally to all of the early ring-shaped transformers. The space occupied was quite disproportionate to the work done. This, together with the difficulty and expense of winding, eventually led to the abandonment of the ring transformer in its original form of an endless jointless iron ring.

One of the best, perhaps the very best, of the early ring-shaped transformers was presented to the public in 1885, by Messrs. Zipernowski and Deri. In this transformer, which electrically was excellent, but mechanically very faulty, the positions of coils and iron were, so to speak, reversed. The primary and secondary coils, both thoroughly insulated being wound into a kind of solid core, which was over-wound with a heavy layer of iron wire. Owing to the shortness and abundant sectional area of the magnetic circuit, this transformer had excellent self-regulation, and owing to the large radiating surface of the iron, the heating effect was considerably reduced. Mechanically, the construction was obviously open to many objections.

About the date of the perfecting of the Zipernowski & Deri transformer, the more modern "block shaped" class of converter commenced to attract attention.

These transformers are of many kinds, no one differing very essentially from another except in details of construction, proportioning, efficiency and workmanship. This class of transformers generally has the coils and iron arranged about as shown in the accompanying drawing, Fig. 18 C.

Thus the coils are entirely surrounded by laminated iron save at their ends, whilst the magnetic circuits are comparatively short and in two directions, the whole apparatus being mechanically simple, and easy to assemble.

Block transformers are those chiefly in use at the present day, and whilst the various makes vary in detail, their general plan is similar. The efficiency of the modern transformer is very high,

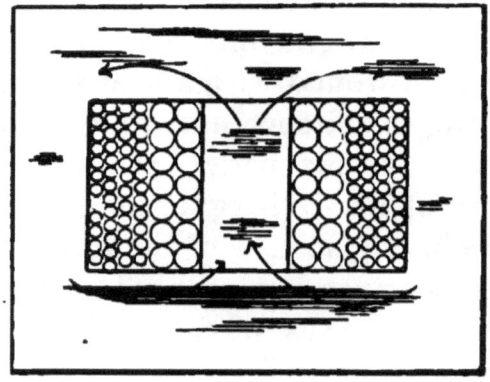

FIG. 18 C.

and its regulation is admirable. It may be taken as an unquestionable fact that more light is being distributed today by means of the converter than by all of the direct current systems combined. There are still some strong supporters of the open magnetic circuit transformers of which the "Ruhmkorff coil" is a type, and by careful construction and proportioning, some remarkable results have been obtained from time to time.

A notable example of this is the English "Hedgehog" transformer of Swinburne, which attracted so much attention some time since.*

As a class however, the open circuit transformers from their very principle never have and never can become generally successful. There are one or two transformers now in use of rather unique design, which show no especial feature of excellence, and which form no link in the chain of improvement which we have attempted to follow, but which are none the less reliable and efficient, though perhaps less so than others of the present day. A good example of these is the Ferranti converter, which is referred to elsewhere.

The chief direction of improvement in transformer construction today, tends first, toward greater efficiency, and secondly toward the bettering and simplifying of case and fixtures, such as fusing devices.

Whilst we can scarcely look for the 105 per cent efficiency, which some printed matter has now and then seemed to hint at, still the transformer, of 1892 has crept up very close indeed to the other side of the 100%.

---

* See Appendix.

CHAPTER V.

TRANSFORMER CONSTRUCTION.

In considering the transformer and its structure from a purely mechanical point of view, the student is at once confronted with a multitude of forms, each possessing its own especial advantages. Few it must be confessed, can be found, which do not show some noticeable defects. In preparing this chapter it has been our endeavor to classify the different steps in the construction, treating them progressively, rather than as a whole.

The general character of the transformer is at once determined by the form of core employed, and in this portion of the apparatus, good mechanical construction, probably does more to influence the excellence of the results than in any other portion of the work.

On pages 74-75 are shown some typical forms of cores, all more or less in use at the present day.

No. 1 is without doubt the most typical and characteristic form of the modern transformer, the "block-shaped" converter of the preceding chapter, A and B are respectively, the iron and the

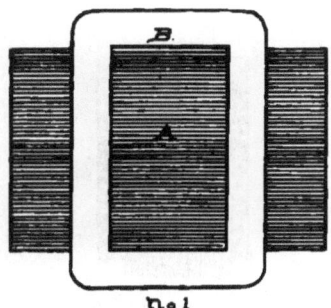

No. 1

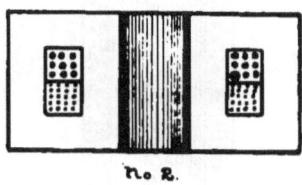

No. 2

Fig. 19.

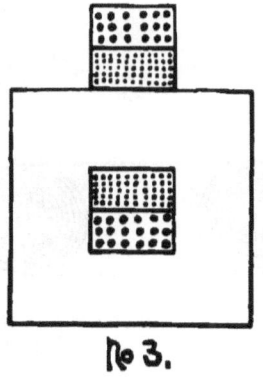

No. 3.

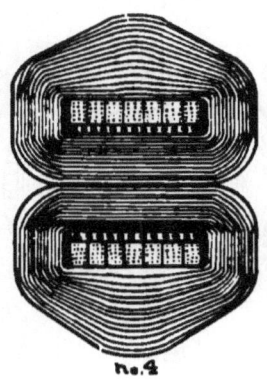

No. 4

TRANSFORMER CONSTRUCTION. 75

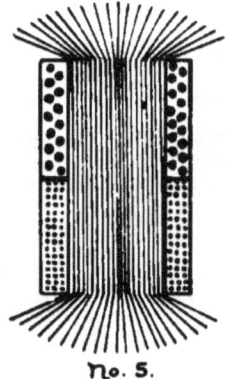
no. 5.

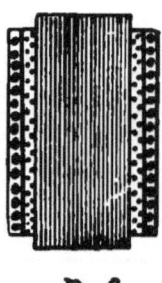
no. 6.

Fig. 20.

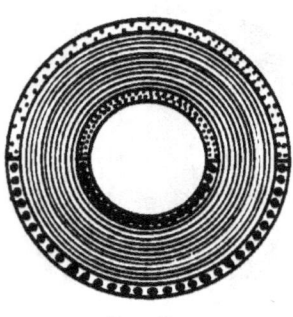
no. 7.

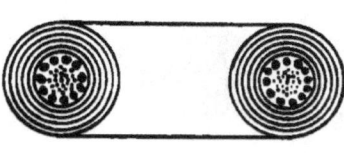
no. 8.

winding. A is laminated, in the direction shown in cut, which is of course, a section. It will be noticed at once that the coils in this form of converter are easy to wind and most compact. Some manufacturers wind the primary and the secondary into two coils of equal internal diameter, placing them upon the iron side by side, others again wind either the primary or secondary of greater internal circumference so that the one may fit snugly over the other; whilst in some instances the two are combined in one solid mass. In this latter case the primary or secondary is first wound upon the mandril and completed, this then in its turn, is overwound with due insulation, upon which is placed the second coil, the secondary generally forming the outer layer. It is obvious that in this form of converter, the iron must be divided at some point to permit of the introduction of the coils.

Very many different methods have been, and, in fact, are still, in use for joining the iron, the object, in all cases, being to obtain as good a contact as possible between the abutting ends, and thus avoid the introduction of unwarranted magnetic resistance resulting from a gap or break in the path of the lines of magnetic force. Some manufacturers prefer to break each plate of the lamination at the same line, thus permitting the two blocks formed by clamping the plates together, to abut as two solid masses, whilst others break the plates at alternate lines, thus overlapping each break with the solid portion of the next sheet above it.

A clear idea of what is meant may be had by glancing at cut 21. In the latter method the iron necessarily has to be built up within the coils, which probably renders the assembling of the transformer somewhat more laborious. In either case a very perfect magnetic joint is secured by proper care, and at this point as in almost every

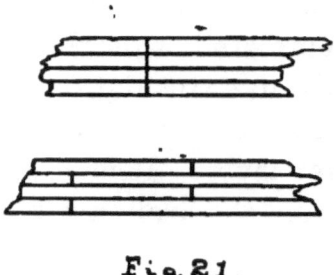

Fig. 21.

other detail of construction, care in the manufacture bears a very important relation to the final efficiency obtained. Cut 22 shows some of the commoner lines at which the cores of block-shaped transformers are broken. In transformers of medium and large size, it is a very common practice to wind several coils, each of primary and secondary.

Cut 2, Fig. 19, shows us another form of transformer in use at the present day. This as well as No. 8, Fig. 20, (which it closely resembles) is a ring-shaped transformer, having the iron (which we can scarcely term a core in this case) on the outside. In transformers of this type, the primary and secondary are formed into a solid internal ring, the two circuits being of course duly separated one from the other. Upon this ring of wire is placed the iron. In the case of type 8, this con-

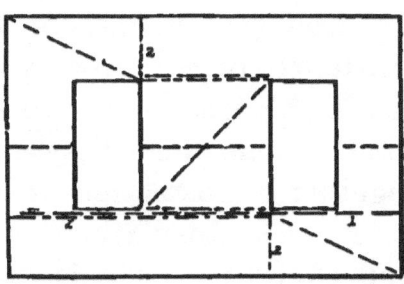

Fig. 22.

sists of iron wire overwound upon the coils by hand, a tedious and expensive operation.

This form of construction is seldom, if ever used. Commonly the iron consists of sheets or plates, as in the block form, and as before, the plates have, of course, to be broken at some line, to permit of the free introduction of the winding.

This form of transformer presents the merit of a short magnetic circuit. The external circumference of the ring being of course greater than the internal, it follows (the plates of iron being of uniform thickness) that the peripheral edges of the plates must be somewhat separated, even though the inner edges be clamped very closely together, and this is a mechanical difficulty to be overcome, although it is condusive to very perfect lamination.

Cut 4, Fig. 19, shows us a section through a transformer of rather unique construction. The iron in this case consists either of bundles of very thin soft hoop iron well bound together, or, in a few isolated instances of soft iron wire. A number of bundles of this iron being brought together parallel, and in the same plane, they are overwound and bound together by insulation at their central portion. Over this insulation is wound the secondary, and over this again is placed the primary, generally in the form of ready wound coils, due insulation being interposed. The soft iron is then turned back and over from each end, the ends of the strips lapping one over the other, till the middle of the bundle is reached, when the last two ends are turned back and made fast; the remaining half of the iron is then turned back similarly in the opposite direction, the iron, when in position, enclosing the coils as in the block transformer.

This design presents some merits, chiefly mechanical, probably the most marked being the ease with which injured coils may be removed and replaced. Where iron wire is used for the core, the device differs somewhat, in that the iron is turned back around the coils in all directions, instead of in two directions only, and in this case, as in most wire-cored converters, the coils are generally circular. Wire-cored transformers of this type are, however, only made in small sizes and for especial purposes. As a type, this general style of converter has been somewhat popular abroad, and has been adapted to very high voltages, large capacities, and step-up purposes.

No. 3, Fig. 19, shows us another form of the popular block-shaped transformer. It will be noticed that the iron only encloses one-half of the wire, hence the converter is, generally speaking, less efficient for the same weight of iron than the usual block form. Transformers of this type are used solely for special and experimental purposes. Its only merit lies in its extreme simplicity.

In cut 7, Fig. 20, we have the modern form of the original Faraday "Perfected Induction Coil," a typical ring-shaped converter. The iron core is of wire, wound upon a form and bound closely together with insulating material, which serves to isolate it from the coils. The primary and secondary are over-wound upon this, necessitating much hand labor. At times one-half of the ring is devoted to the primary winding and the other to the secondary, as shown in the illustration. Sometimes the two circuits are divided into a number of short coils, placed alternately, and again one coil is wound over the other. Transformers of this type are not common in practice, as they present numerous disadvantages, not the least of which is in construction. They have been used, however, in connection with series lighting, and for special converters of small capacity.

Types of the open circuit transformer are shown in Fig. 20, Nos. 5 and 6. The core is almost invariably of wire, straight and non-continuous, the magnetic circuit being completed through the air. No. 6 is the popular form of the early Ruhmkorff coil, already mentioned in previous chapters. Its general plan of construction is so simple as to need no comment, more especially as it bears no especial relation to electric lighting, although in the early days it was used in series distribution. The same modifications of winding are used in this, as in the ring transformer described as No. 7. Although quite typical and very useful for many purposes, transformers of this class may safely be put aside as essentially inefficient.

Cut No. 5 illustrates the general plan of construction of the so-called "Hedgehog" transformer. Notwithstanding its similarity to No. 6 this converter may be considered as a type, its characteristic being the manner in which the ends of the core are finished. The iron wire is permitted to extend considerably beyond the coils, the wires being bent into a radiating form, so that each individual wire is separated from its neighbors, the whole having the general appearance of a brush. This construction serves to equally disseminate the lines of force through the surrounding space, and results, according to the inventor, in especial efficiency. This theory deserves attention, and is the subject matter of an appendix.

One of the most important points in transformer construction is the selection of the iron. Generally speaking, it is safe to say that the selection of the iron is one of the most important preliminaries of transformer construction. Not only is the purity of the iron important, but also its texture, fibre and malleability. Iron of great purity should be selected. In other words, iron that is as free as possible from carbon, phosphorus, silicon, etc. This may be determined by chemical analysis. A high quality of Swedish soft sheet is probably as good as anything that can be selected, although some of the very high grade English irons are excellent. The metal should be as soft and uniform as possible, that is, it should be thoroughly well annealed. The iron which can be bent the greatest number of times without breaking may, as a rule, be safely selected as the best, provided the analyses have proved it to be pure, which, with rare exceptions, will be found to be the case.

Such an iron should have rather high tensile strength, and should show a high per cent of elongation before breaking. Recent practice has pointed towards certain grades of low steel, as possessing many good qualities for transformer construction, but suitable steel is probably more difficult to obtain than is good iron.

The iron being selected, the next step in the manufacture is the punching of the plates (we are dealing with the now almost universal "block" type of converter). This is accomplished upon a power stamp or press, by means of dies. The sheet metal is placed upon the lower die, which is held stationary, whilst the upper die is brought down upon it, punching out the requisite shape. Where both an outer shape, and a hole, are to be cut out, it is a common practice to accomplish the operation by a single blow, instead of, as formerly, using two dies and presses. This is done by using what is termed a "leading" die, one form of which is shown in Fig. 23. There are many modifications of this plan, and it is used in the manufacture, from sheet metal, of many hundreds of articles. There is almost always some automatic device for freeing the metal after punching. A good set of dies should leave only a very light burr at the edges of the punchings, whilst poor dies will leave a heavy one.

The plates being punched, the next step is the

building up of the core. The direction of the lamination is already understood by the reader, it having been dealt with in Chapter II, when discussing Eddy currents. If the transformer core is of the class which has abutting edges at the break, then the iron is built up quite independent of the coils. If of the class which has overlapping edges, then it is built up within the coils. In either case the general plan is the same.

As has been stated, next each sheet of iron must be placed a layer of insulating material. Only a

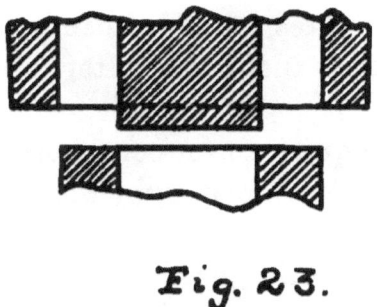

Fig. 23.

very low insulation is needed, for the potential of Foucault currents is invariably low. Practice differs as to the best and easiest method of accomplishing this end. Some makers shellac or enamel the plates, and then bake them in an oven to set the coating. Others rust, or oxidize the plates, rust being an insulator, or, rather, a semi-conductor of sufficiently high resistance for this purpose. Probably, however, thin tissue paper is more universally used for this purpose than any other insulating material. This paper is sometimes glued or pasted to the iron, but more commonly laid in loosely, being cut to the required shape and size, as are the iron plates.

In building up the plates are laid within a rack or former, which they fit, or if there are holes in the plates, as there commonly are, for the binding bolts which are frequently used to hold the block together, then the plates are laid over pins, which correspond to these holes, or even over the bolts themselves. The necessary rack, or pins, being provided, the plates are placed in position, a layer of paper or other insulation alternating with each layer of iron.

When the requisite number of plates have been placed, the whole pile is subjected to heavy pressure, thereby reducing it as nearly as possible to the nature of a solid mass. After this operation it is tightly bound together, either by binding bolts passing through all of the plates, or by clamps on the outside. Where the binding bolts are used they should be surrounded by an insulating bushing, interposed between them and the core, and should have a washer of insulating material under both the head and the nut.

The iron being now ready for assembling, we may turn our attention to the winding and insulation of the coils. This is a comparatively simple matter, and is carried out in the same general manner, whether the coils (primary and secondary) are made up independently, or together, to fit over one another, or to be placed side by side.

Insulation is, generally speaking, the most vital consideration, but this important question will be considered later.

The coils are wound in a lathe, upon a former or mandril, having perpendicular ends like a spool, to hold the wire on, and to determine the width of the coil. One of these ends is removable, to permit of the coils being taken off when completed. Double cotton covered wire is used, generally spoken of as D. C. C. wire. One end of the wire, to be wound up, being made fast, the "former," or mandril, is set in motion in the lathe, the wire, as it is wound on, being drawn through tension pulleys to insure its being wound on straight and level. The wire is watched and guided at the mandril, to insure its being wound on smoothly and progressively. When a layer is completed it is quite customary to give the surface a thin coat of shellac before winding the wire back over itself. In winding the primary coils of large high voltage converters, it is quite customary to insert some kind of insulating material, in addition to the shellac, between each layer, to prevent the difference of potential between the layers from short-circuiting, inner and outer turns.

When the requisite number of turns have been wound onto the former, the wire is cut off and the coil removed, it having previously been tied firmly together at four or five points, by pieces of twine or tape placed on the mandril before the winding began. With the exception of the care required in insulating, primary coils are far less difficult to wind than secondary, owing to the fact that the wire is much finer and more manageable. As the heavy wire necessary for the secondaries of large capacity, low potential transformers is often most unmanageable, amounting almost to rod, copper tape is at times substituted for this heavy wire, with good results. The most popular plan, however, is to wind a number of wires in multiple, or even to place several coils in multiple.

Too much care cannot be exercised to prevent any portion of wire, however small, which is bared of its insulation, from being wound on. Such a spot, if present, would almost certainly make trouble eventually by coming in contact with other bare metal, and causing a short circuit. When such a point is discovered in the wire it should be carefully wrapped with adhesive rubber tape and shellaced before being wound on. If the rubber tape is too bulky, as it frequently will prove to be, then narrow silk tape or ribbon may be used, it being made to adhere by means of shellac. If it becomes necessary, for any purpose, to make a joint in the wire of an incompleted coil, it should always be soldered, as well as twisted, and should be over-wound with tape as already described. Above all things, great care should always be used to see that there are no sharp edges or projections at the joint, which might chafe through the insulation and expose the metal.

In the case of high potential transformers of large capacity, it is a common and excellent practice to wind the primary in several separate and distinct coils, which can afterwards be connected in series, thus preventing any two turns of wire, with a considerable difference of potential, from crossing. If they did so the insulation would almost certainly chafe through and a short circuit result

All of the above-mentioned contingencies having received due and proper attention, the coil may be removed from the "former" without fear of future trouble.

The next operation is the taping of the coils. An ordinary cotton tape, from an inch to an inch and a half in width, is commonly used for this purpose. One end of the tape is made fast to the coil with melted shellac, the covering then being wound on and over the coil progressively. When the coil is intended to carry, or be brought into close proximity to high pressure, it is preferable to insert a layer of insulating material below the outer covering of tape, mica is probably the best substance for this purpose. It should be put on over a layer of heavy shellac, and should have another coating of thin shellac before the tape is applied. Mica is very brittle, and great care should be exercised to prevent its breaking and chipping off at corners and curves. Other substances are frequently used instead of mica, such as asbestos paper, vulcanized fiber, shellaced paper, etc., but whilst far more easy to manipulate than mica, they are inferior to it, owing to the fact that they absorb moisture more or less rapidly, which speedily destroys their insulating qualities until they are again dry.

No mention has been made as yet of what is done with ends of primary and secondary coils. They are, of course, brought out through the tape, and this should be done in such a way, that their positions may be comparatively remote from one another, since contact between the two terminal wires, especially of the primary coil, even though well insulated, would be most dangerous, and might at any time result in a short circuit and a burn-out. The terminal wires should be carefully taped and shellaced, both for the sake of the insulation, and the protection which the covering affords to the wire. It is a common plan to bring out the ends of the primary coil at one end of the transformer, and the secondaries at the other. A sufficient length of wire is left to permit of making the necessary connections to the binding posts, or terminal blocks, which will be presently mentioned.

On completion, the coils are given a final coat of shellac, after which they are sometimes baked in an oven, to thoroughly dry them out and harden the shellac, but this is not always necessary, since the completed transformer is generally baked as a whole before being enclosed within its case.

The finished coils are next placed upon the iron in their proper position, where they should fit closely, due insulation (dependent upon the pressure for which they are intended) being interposed. Mica, rubber, vulcanite, fibre, asbestos and wood, are the substances commonly used for insulation in transformers.

There is often a space between iron and windings, at the bend at each end of the coils, caused by the coils not making a square turn and lying flat upon the core. A wooden block, or wedge, is frequently inserted at these two points, to render the whole solid. Teak, or failing that, a very resinous pine is probably the best wood as regards insulating qualities, for use in transformer construction. The coils being in position and the two portions of the core being brought together and firmly fastened, either by clamping screws or bolts, the transformer is complete and ready for baking, to which process it is at once submitted, being placed in an oven which is at a temperature as high as the insulation will bear without injury. This thoroughly dispels all moisture and sets the shellac hard, so that it resembles lacquer. On being removed from the oven, and cooled, the transformer only lacks testing, casing and connecting, to be ready for commercial use.

A thin cast-iron box or case is used to enclose the transformer. Although this case varies greatly in details of construction, it is always water-proof and usually has some provision for ventilation, a ring bolt for lifting, lugs and holes for lag screws, by which it may be fastened up. Terminal blocks are set into it, but insulated very carefully from it, and these blocks should be so placed as to preclude any possibility of a short circuit between them. The two ends of the primary and secondary coils are fastened to these blocks from the inside, and the ends of the leading-in wires from the main circuit to the primary, and the leading-out wires from the secondary to the lamps, are fastened into these terminals by means of set screws when the transformer is put into use. The greatest care should be exercised to prevent the necessity of crossing either the primary or secondary inside leads to get them to their binding posts. This is especially important in connection with the primary, as, if they cross, a short circuit is very likely to result, if the insulation of the leads becomes chafed at the point of contact.

It is a common practice to have an internal arrangement whereby fuses, which can readily be replaced, are interposed between the terminal wires of the coils and the binding posts, sometimes in connection with the primary coil only and sometimes with both. Such arrangements are generally mounted on porcelain, and placed within a separate compartment of the transformer case, which can readily be opened to replace a burned out fuse. Secondary transformer fuses, though often found, are, generally speaking, uncalled for. Although, till recently, fuses have almost invariably been placed within the transformer, it is probably better to have them outside instead, in an accessible position, and this plan is rapidly finding favor.

Many manufacturers have their transformer cases so constructed that the fuses cannot be reached without breaking, or opening the circuit. As a safety device this is excellent, as many accidents have occured from carelessness in putting in new fuses on "live" circuits.

One of the most radical, and marked improvements in modern transformer construction, is the introduction of oil insulation, the most perfect and efficient insulation known for this class of work. Where this is used, the entire transformer is submerged in a heavy low grade oil, specially prepared for the purpose, and closely resembling a low grade of cylinder oil. This dense fluid speedily penetrates every crevice of the transformer, and becomes a perfect insulating film at every point of danger, precluding besides any possibility of moisture.

If a puncture or short circuit occurs at any point, from coil to coil, or from coils to iron, the oil at once penetrates and re-insulates the crevice, rendering the point where the discharge took place as perfect as before, thus it may be said that transformers, with oil insulation, are self-repairing.

Converter testing in practice, is a comparatively simple matter. The points to determine, are the thorough insulation of the coils from one another, and from the iron and casing. The fitness of the transformer to carry its normal load indefinitely, its ability to bear its normal voltage, or a greater pressure, without breaking down; and the ratio of conversion, to make sure that the right number of turns are present, or that the voltage of the secondary is correct, with normal primary pressure.

To test the insulation, a magneto capable of ringing through a high resistance is applied. The contact points of the magneto should be applied first to the end of the primary and secondary coils, then to primary and iron, and to secondary and iron. If the insulation is perfect no ring will be obtained. Some makers test by attaching primary to one side and secondary to the other side of a very high potential circuit, of say five or six thousand volts. The insulation should stand this without breaking down.

To test the ability of the transformer to carry normal load, and bear normal pressure, it is only necessary to place it on a standard primary pressure, load it to its full capacity of lamps and leave it there for some hours, testing the voltage on primary and secondary at the end of the run to determine that the ratio of conversion is correct.

The converter having stood these tests, is finished and ready for shipment and service.

We have endeavored to carry the reader with us through the path of construction in its main features, from the transformer's component parts to its completion, and it now only remains to us, to add a few points of great importance, which should be borne constantly in mind, to obtain the best results.

1st. Never permit two wires with any great difference of potential to cross each other. Such construction invites a short circuit.

2d. Allow ample sectional area of copper, (800 to 1000 circular mills per ampere,) otherwise the converter will run too warm, and give poor results.

3d. Better use too much than too little insulation.

4th. Do not attempt to work the iron at too

high a point of saturation, that is at too high an induction, it is not economical.

5th. Clamp the iron and bind the coils rigidly, if this is not done the transformer may hum.

6th. Ventilation is an excellent thing, but moisture is dangerous and distructive.

7th. Keep the primary and secondary terminals well apart.

8th. Do not let any iron chips, or filings, get into the transformer, they will make their way through almost any insulation by magnetic propulsion, and perhaps cause a short circuit.

CHAPTER VI.

THE TRANSFORMER IN SERVICE.

If the theory of the transformer is in some of its details somewhat intricate, its application is simple and easy to a degree. We have in the foregoing pages treated here and there of various matters which have direct bearing on the subject of this chapter, and we therefore shall here limit the discussion to various important points of detail, rather than to any general plans of arrangements of circuits, which are undoubtedly already well understood.

The question of the location of a transformer is an important one, and is scarcely amenable to any general rules which could be formulated.

Local custom has much to do with the positions selected for installing converters. In Europe it is the general practice to install them *within* the building whose lighting service they are to maintain, and for this reason the transformers manufactured in Europe are not, as a rule, enclosed within waterproof cases. With us, however, it is the custom to install converters on the *outside* of buildings, or in some position in the open air, and this is generally demanded by insurance requirements.

There is an opportunity for the exercise of much judgment in selecting the capacity of the transformer required to do certain work. It should always be borne in mind that the nearer a transformer operates to full load the greater its efficiency will be, but that an overload is a bad thing, endangering the transformer and causing poor regulation. If the transformer is to operate on a dwelling house circuit, then the number of lamps upon its secondary will be considerably in excess of the actual capacity required, for there will seldom, if ever, be an occasion when all of the lamps will be in service at the same time. In a case of this kind a close estimate of the number of lamps likely to be in use at one time will represent the capacity required. If a store, hall, or factory is to be lighted it is probable that all of the lights will be in use at the same time and, therefore, the total number of lights installed will represent the size of converter required.

Large transformers are more efficient than small ones, therefore they should always be used when possible. It frequently happens that a number of stores or dwelling houses requiring light are near together. In a case of this kind one transformer can be made to do the work for the whole number, it being installed as nearly as possible at the centre of consumption, and its secondary being divided and carried to the various installations, as separate secondary circuits. This practice, if generally followed, will result in economy, though, of course, the secondary must not be carried through any considerable distance, or the C_2R loss will become excessive and wasteful.

There is an opportunity also for the exercise of much judgment in determining the best secondary potential to be used. If the total secondary circuit is short, then 50 to 52 volts is the best pressure to use, for, though the amount of copper required will be considerable, lamps of this potential are somewhat more efficient. If the length of secondary wiring be considerable, then it is generally preferable to use a pressure of 100 to 104 volts, because of the resultant saving in wire, and the reduction of C_2R loss. Where a large building is to be wired and the total length of secondary wiring is great, it is a frequent and excellent practice to bank two transformers and do the wiring on the three-wire plan. Directions for doing this will be found later in this chapter.

Where the number of lights to be carried on one circuit is greater than the capacity of a single converter, a number may be arranged in multiple, as shown in Fig. 24. When this is done however, the transformers used, should always be of the same *capacity*, and the same *style*. It must be

TRANSFORMER IN SERVICE.

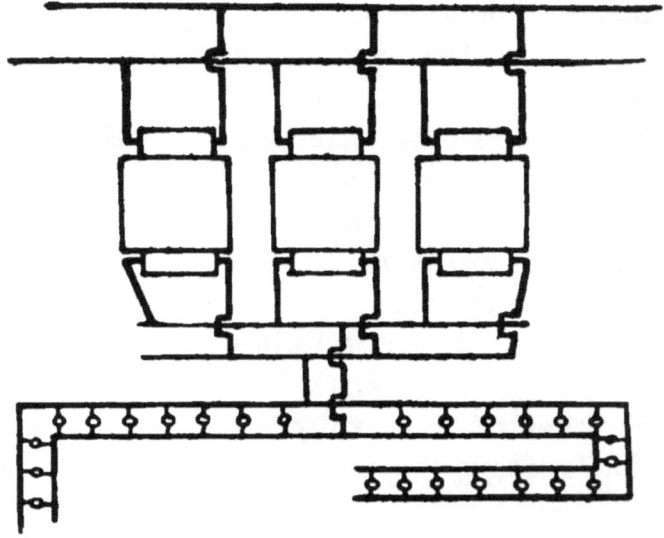

Fig. 24.

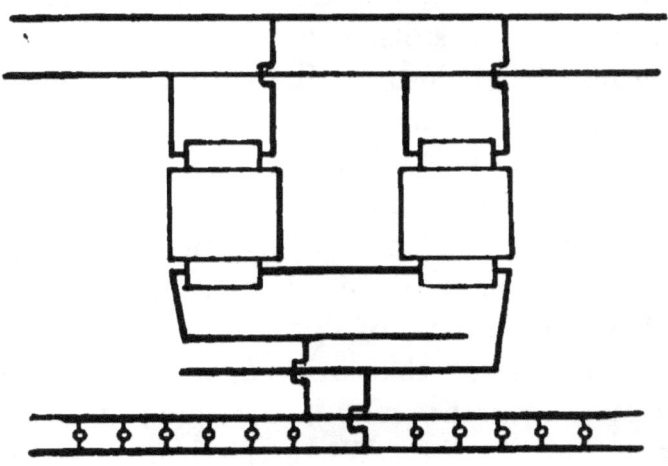

Fig. 25.

borne in mind, that if the primary fuse of one of several banked transformers "blows" then the secondary of that transformer becomes a load upon the rest of the bank, in *addition* to the portion of the load it was carrying. Under this extra strain, the fuses of the remaining transformers, will probably blow also, and darkness and confusion result.

For this reason it is generally preferable to divide all circuits when possible, using single transformers, so that all of the lamps in a building, may not be dependent on a single source. When banking is unavoidable, it is an excellent plan to insert a large main circuit fuse on the united secondary of the whole bank, calculated to blow before the individual transformer fuses, thus, if all of the lamps go out, they can be re-lighted by the insertion of a single fuse instead of one in each transformer. It is of course understood that there are always a number of sub-circuit fuses between the lamps and the bank.

At times it becomes necessary to so bank transformers, as to get an increased potential. For example, a building may be ready wired on a basis of 100 volts, the lamps being made for 100 volts also, whilst the only transformers available are designed to give a secondary potential of 50 volts. In this case the transformer must be banked with two secondaries in series, as shown in Fig. 25.

It is well to remember that *when two or more transformers are connected with their secondaries in multiple*, their united output is equal to *the sum of their rated current at their normal potential.*

When banked with their *secondaries in series*, their output is equal to *the sum of their rated potentials, with current equal to the rating of one transformer.* (All of the converters, of course, being of the same capacity.)

To bank transformers to operate on the three-wire plan they should be arranged with their primaries in multiple, as is usual, and their secondaries in series, with the neutral, middle, or third wire taken off between them, as shown in **Fig. 26.**

This is generally done with 100 volt transformers, and saves much wire, the greater portion of the energy being distributed at 200 volts, and the lamps (which are 100 volts) burning two in series, except when more are in use on one side than on the other, in which case the middle wire takes care of the difference. The two outside wires are to be figured upon on the basis of 200 volts, and the middle wire should have from one-half to one-third of the carrying capacity of the outside wires. The lamps must, of course, be distributed about equally between the two sides. The main fuse on the middle wire must have a capacity in *excess* of its calculated load, for if it burns out many of the lamps on one side will, as a rule, be destroyed by

102　　　　　　TRANSFORMERS.

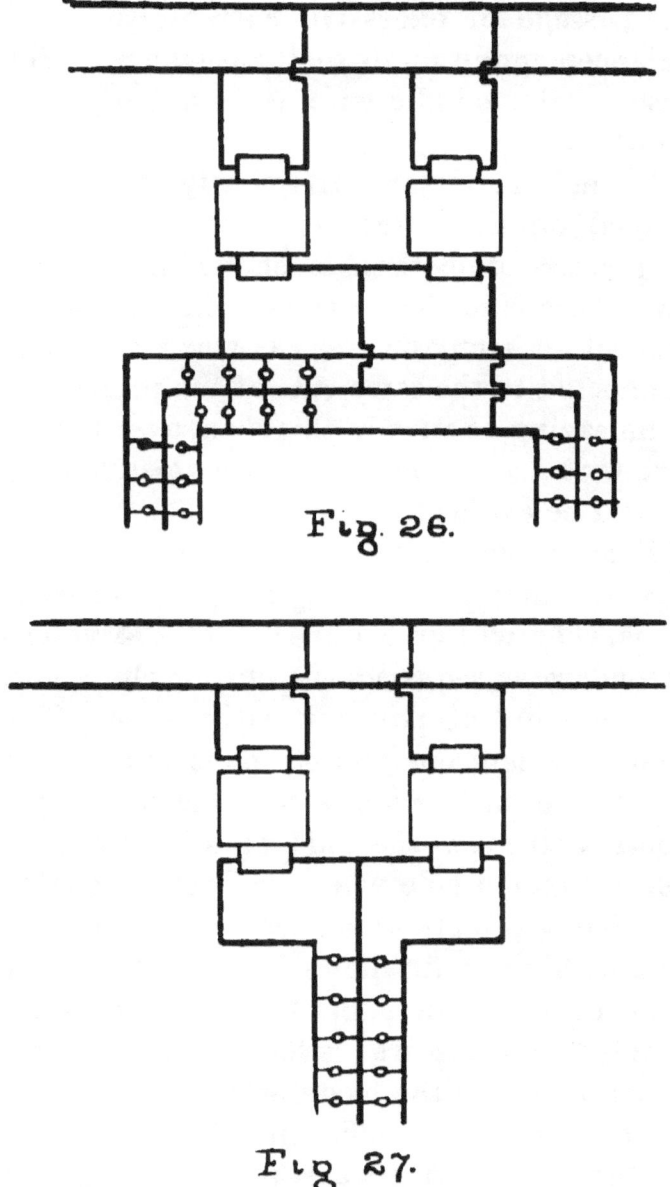

Fig. 26.

Fig. 27.

the passage of excessive current, due to the unbalanced condition of their resistance. The main fuse on the middle wire is frequently altogether omitted.

Transformers are frequently incorrectly arranged on the three-wire plan. A common but *very incorrect* method is shown in Fig. 27. The error here is in the primary connections, opposite, instead of similar wires (as when correct), being connected to the same side of the primary. There is no saving of wire in this plan, over the ordinary two-wire arrangement. It has no advantages, and should be avoided.

Some commercial transformers are so constructed that by changing simple internal connections, they will give either 52 or 104 volts at the secondary as required, the output representing the same amount of power in either case. This plan is an excellent one and of great convenience.

Transformers should be placed, when out of doors, either on the side of a building, or, still better, upon a pole when available, the latter position being preferable on account of the vibration. Jar, within certain limits, is an excellent thing for transformers, and should be sought rather than avoided, since, as we have seen in Chapter II, it tends to reduce the hysteresis loss, and to render the converter more efficient.

Many transformers are provided with hooks, or

a kind of elbow at their upper end, by means of which they can be suspended from the arm of a pole, or from a cleat attached to a wall. This is a great convenience and saves much labor. Generally speaking, however, they are fastened up by means of lag screws.

If it can be avoided, transformers should never be placed against metal, such as a tin roof or metal shingled wall, as there is always danger, in a case of this kind, of a "ground" on one side of the line and consequent trouble. It is also a bad practice to lay converters on their sides or backs, in positions where they are liable to be exposed to wet. This is frequently done, but is none the less to be avoided. The upper end of the case is designed to withstand rain, while the remaining portions of the box are not always absolutely waterproof in case of a heavy downpour, and if water penetrates within, trouble is more than likely to result.

In some localities, where violent electrical storms are prevalent, notably the Southern countries, considerable damage is at times caused to transformers by lightning. In such cases it is not necessary that a lightning discharge should "strike" the converter or its wires, though this at times occurs. Often the trouble results from a *secondary current*, induced by the lightning, generally in the main primary circuit. The effect which this has upon the converter varies greatly with the conditions.

At times it simply blows the primary fuses, at others it melts a portion of the primary coil.* More commonly, however, it punctures the insulation, striking through either to the secondary coil. the iron, or both, at times leaving no evidence of the mischief which has been wrought; it is this last contingency that is most to be dreaded, for it may contain an element of danger.

The primary having a path to the secondary on one side may, if a "ground" occurs upon the other, send a dangerous shock through the secondary, provided anyone should come in contact with a portion of the secondary circuit in such a manner as to connect it to the ground through themselves. Such a combination of circumstances is of necessity rare, however, and there is probably only one authentic case of disaster on record, resulting from this state of affairs.

It is in connection with punctures, due to lightning, etc., that the system of oil insulation is of greatest value, for every puncture is re-insulated by the oil as soon as made, leaving the transformer practically as perfect as before, Another safety device, which has been adopted in some transformers, in connection with lightning discharges, calculated to prevent injury to the coils, is the grounding of the iron core. This is accom-

---

* This latter result is rare, but has been observed. When the fuses blow, they must have drawn an arc, which maintained the circuit.

plished by connecting the laminated plates either to water or damp earth by means of a copper wire. Since lightning invariably seeks the shortest path to the earth,* it will, when the core is grounded, strike through the insulation to the lamination, and thence go directly to earth, when if oil insulation be present it will immediately re-insulate the puncture, leaving the transformer in perfect working condition. When the core is grounded in this way it must, of course, be strictly insulated from the case. It is naturally understood that the potential of a lightning discharge is practically irresistable.

The primary fuses in the transformer, and in fact everywhere, should have a rubber covering over the lead and tin wire of which they are made. Primary fuses are now constructed in this way. The rubber covering prevents the flash of a burning fuse from reaching beyond its proper limits, and also tends to prevent the burning fuse from establishing an arc between its terminals.

When not enclosed within the transformer, primary fuses should be placed out of doors, within boxes manufactured for their reception, and in a readily accessible position, conveniently near the transformer. Both the main and sub-circuit secondary fuses should be placed indoors, in a position where they can be readily reached.

---

* The shortest *electrical* path, *i. e.*, the path of least resistance.

In very long distance distribution it has sometimes been found advantageous to introduce *step-up* transformers, to raise the output of the dynamo to a yet higher pressure (5,000, and even 10,000 volts), at which potential it is carried very great distances with remarkably little loss, and on relatively smaller wire, to the distributing point, where it is again reduced to customary distributing pressure, and carried over ordinary circuits to the house transformers, the energy thus being transformed three times before reaching the lamps. This system is valuable in that it renders possible the utilization of large water-powers, remote from points of usefulness. The plan is quite popular in Europe, but much less so in this country. Of course the great difficulty to be met lies in providing the insulation requisite for such high voltages. Further information relative to this subject will be found in an appendix.

It will sometimes be noticed that transformers emit a humming noise, which is occasionally so loud as to be annoying. This will be especially noticeable if the converters are carrying an overload. The noise is caused by the rapid vibration of the iron plates of the lamination; and sometimes even by the coils themselves, resulting from the rapid attraction and repulsion of the iron plates of the lamination, and also of the individual turns of the coils. The more tightly the plates are clamped together, and the more closely the coils are bound, the less the transformer will hum, in fact, well constructed transformers seldom hum noticeably, even upon considerable overloads.

There are several very important points in connection with the installation of transformers which should be carefully borne in mind. An endeavor is here made to condense these into a few terse and clear sentences.

1. Remember that the *larger wires are the secondary connections and go to the lamps.*

2. Do not put in the fuses till the transformer is fixed in position.

3. *Rubber covered fuses should be used on the primary circuit.*

4. Do not handle *live primary wires without rubber gloves.* It is dangerous.

5. Try to use *only one hand when working about live primary wires.* It insures safety.

6. *Locate all fuses where they can be readily reached.*

7. Place the main primary fuse out of doors if possible, when not enclosed within the transformer.

8. *Abide by the Underwriters' Rules.* (See Appendix.)

9. Avoid making all lights in a large building dependent on a single bank of transformers. *Divide your circuits.*

10. In banking *use transformers of the same type and capacity.*

11. When banking *connect similar leads of the transformer to the same side of the primary line.*

12. A slight amount of *jar renders transformers more efficient.*

13. *Large transformers are more efficient than small ones.*

14. Do not use larger transformers than necessary. *Transformers are most efficient on full load.*

CHAPTER VII.

COMMERCIAL TRANSFORMERS.—FERRANTI.—NATIONAL.—SLATTERY.—STANLEY.—THOMSON-HOUSTON.—WESTINGHOUSE.

It has been thought best to devote a brief concluding chapter to a few of the commercial transformers which are to be found upon the market at the present time. The transformers mentioned in the following pages do not represent all that are for sale at the present day by any means. They have been selected either because of their prominence and popularity, or because novel and typical. They are all, with one exception, of American manufacture.

The Ferranti Transformer is the only converter of foreign manufacture of which mention is to be made in this chapter. It has been selected as typical, and a good example of European practice.

It will be noticed by glancing at the accompanying illustration, Fig. 28, that this transformer is not enclosed within a water-proof case, as is customary with converters of American manufacture. Such a case is not needed where transformers are to be

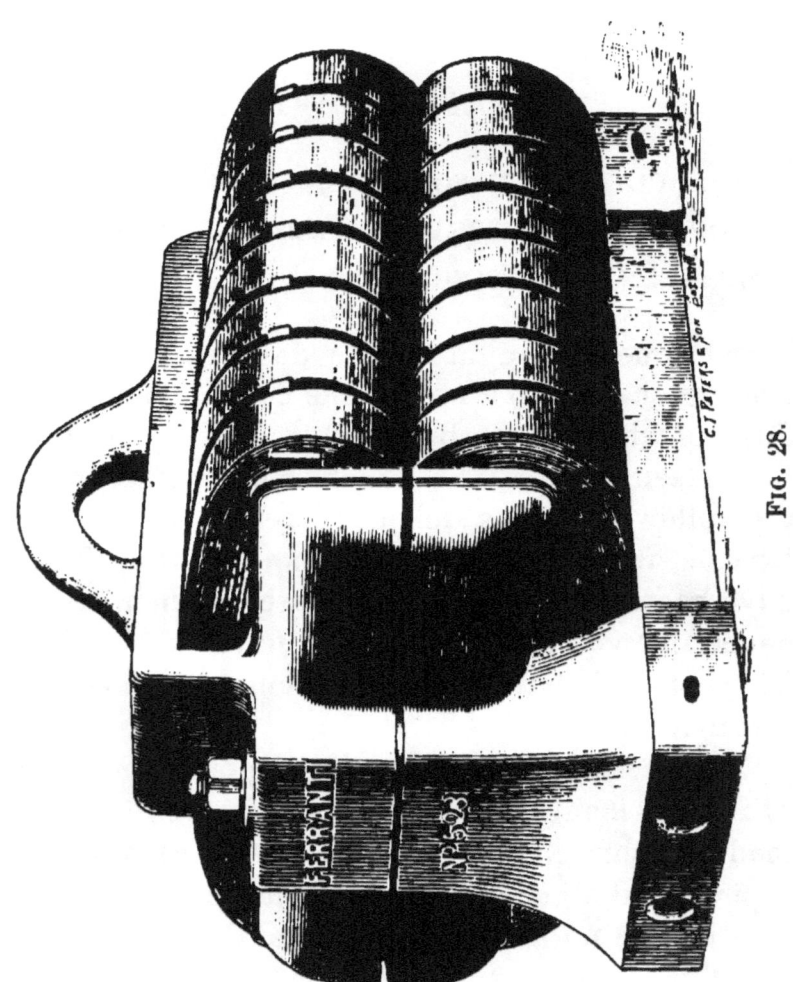

Fig. 28.

installed within buildings, as is usual in Europe. This transformer is typical, being built on the plan of type 4, Fig. 19, Chapter V. The frame which holds and supports the actual converter is of cast-iron, and is so constructed as to provide for standing the transformer upon the floor, a basement or dry cellar generally being the location selected.

The primary and secondary terminals are at opposite ends of the base, and are so constructed that they cannot be tampered with, or the wires loosened with an ordinary screwdriver. The terminals are thoroughly insulated from the frame by means of sulphur and glass insulation, poured, while in a molten state, into the space between the frame and each terminal block. No fuses are contained within the converter, as it is intended that they should be installed separately.

The iron used in the construction of these transformers is extra soft Swedish sheet, and is unusually thin. As it is not clamped together, however, the iron is quite apt to be loose, and the transformers have a tendency to hum. The general plan of construction follows closely that described under the head of No. 4, Chapter V, in fact, this is probably the only commercial transformer of that type. Ferranti transformers have been constructed for service on very high potentials, 2400 volts being the commonest primary pressure, while large central distributing trans-

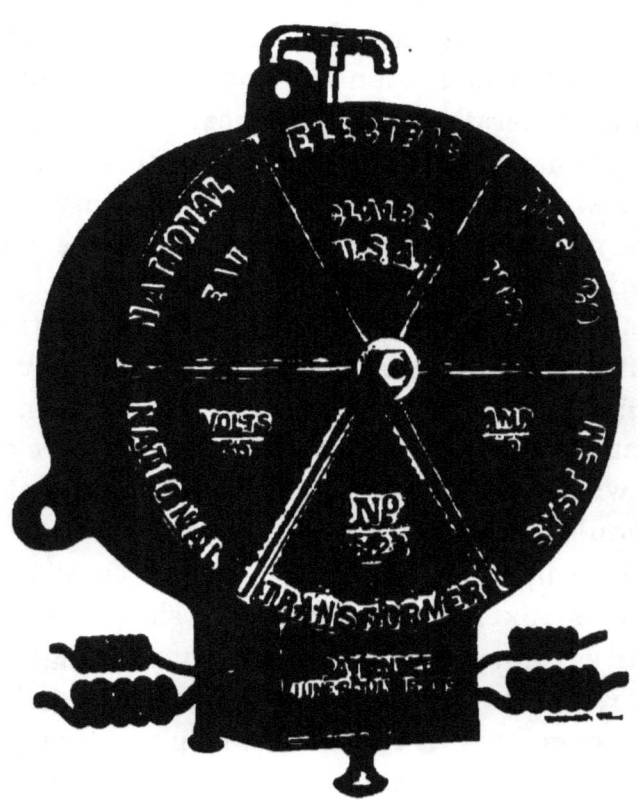

Fig. 20.

Fig. 30.

formers have been constructed for, and successfully operated upon, 10,000 volts.*

This is the class of transformer used in connection with distribution from the celebrated Deptford Station, near London. The chief merits claimed for the Ferranti converter are high insulation and the ease with which they can be repaired.

The National Transformer is manufactured by the National Electric Manufacturing Co., of Eau Clairs, Wisconsin. Its general appearance is clearly shown by Fig. 29. The transformer is of the ring type, described in Chapter V as No. 2. The entire winding is surrounded by iron, all of the wire in the transformer thus being active, and high efficiency is claimed as the result. The true character of the converter may be better understood by referring to Fig. 30.

A novel feature is the fuse and connection box, which is on the lower side of the case. The opening of the fuse box door simultaneously breaks the connection between the primary wires and the fuse contacts, thus rendering it possible for the primary fuses to be quickly replaced without danger of shocks or short circuits, the transformer and its secondary circuits being inactive when the fuse box is open. This safety device is clearly shown by Fig. 31. The case is so constructed as to be

---

* See Appendix.

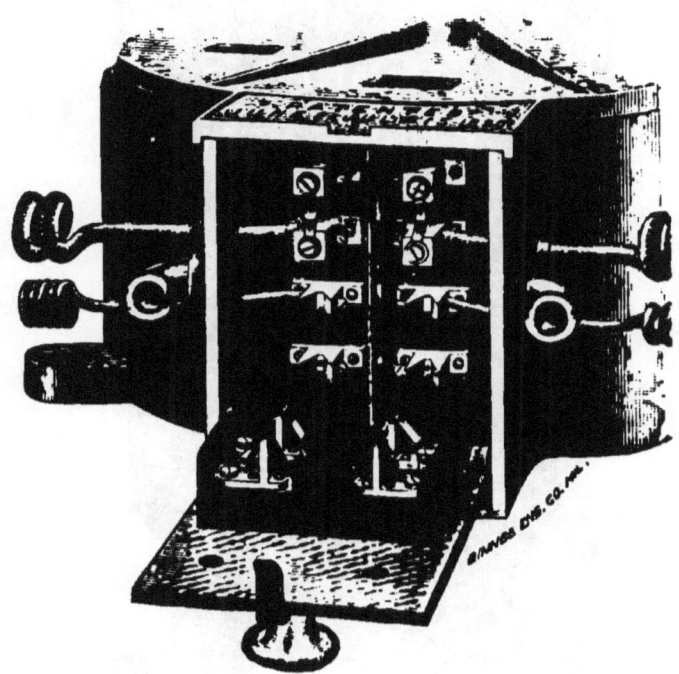

Fig. 31.

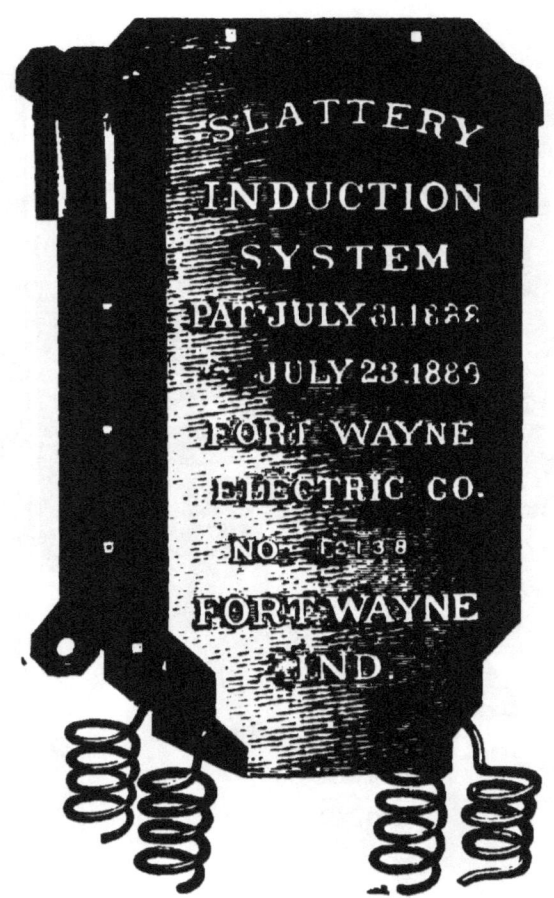

Fig. 32.

waterproof, and ample provision is made for ventilation.

These converters are manufactured in sizes ranging in capacity from two to one hundred 16-candle power lights. The safety device is not furnished for transformers of less than five, or more than fifty light capacity. The merits claimed are high efficiency, safety, convenience, and close regulation.

The Slattery Transformer. This converter is manufactured by the Fort Wayne Electric Co., of Fort Wayne, Indiana. It is a block-shaped transformer, belonging to the class described as No. 1, Chapter V. Its general character is clearly shown by the accompanying cut, Fig. 32.

Both the primary and secondary leads enter at the lower end of the case, being kept well apart, however. The case is cast in two nearly equal pieces, and is bolted together, as shown in the illustration. Careful provision is made for proper ventilation, thus insuring cool running.

The ratio of transformation adopted is either 20 or 10 to 1, 50 or 100 volts being the popular secondary potentials. Efficiency, simplicity, safety and careful construction are the merits claimed for this transformer. It is widely used and generally gives excellent results.

The Stanley Transformer is manufactured by the Stanley Electric Manufacturing Co., of Pittsfield,

COMMERCIAL TRANSFORMERS. 119

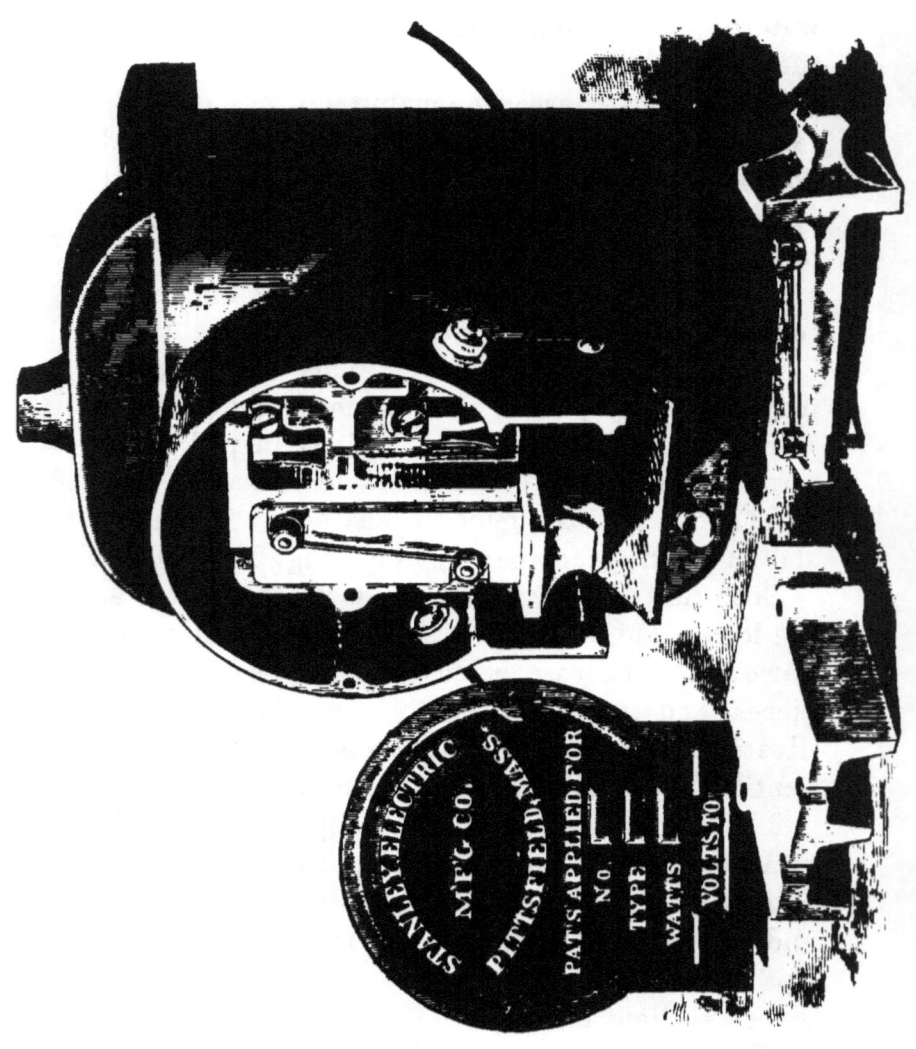

Mass. This company devotes itself almost exclusively to the manufacture of electrical converters. The construction of the Stanley transformer is clearly shown in the accompanying illustration, Fig. 33. It is here shown with the fuse and connection box open, showing the construction. No secondary fuses are used. The primary fuses are mounted on movable porcelain blocks, one of which is shown removed from its position.

In inserting new fuses it is only necessary to withdraw the blocks from their connections, which breaks the circuit, after which a new fuse can be set on the block without difficulty or danger. It is customary for a station to have a number of extra plugs, that there may be no delay in inserting a new fuse. The same size of plug fits all different sizes of transformers up to 100 lights. The front plate of the fuse box is held in place by two screws, and the lower flap is held up by a thumb-screw, in contact with a lip cast onto the front plate. The two fuses are entirely separated by a porcelain partition in the fuse box, thus preventing any possibility of a short circuit from one to the other. The secondary connections are made at the lower end of the transformer.

In replacing fuses the front plate is not removed, but the lower flap, or door, is opened by loosening the thumb-screw. The transformer is provided at the back with a hook or elbow, forming part of the case, thus rendering it easily placed in position. It is only necessary to hoist the transformer up by the eyebolt and hook it over the cross-arm of the pole, or a timber on the side of a building. The merits claimed for this transformer are high efficiency, close regulation, small leakage, convenience and safety.

The Thomson-Houston Transformer. This converter is manufactured by the General Electric Co., at their factory in Lynn, Mass. It is of the block form, but embraces many novel features. A clear idea of the general character of this converter may be obtained by glancing at Fig. 34. The leading-in wires for both primary and secondary are at the top of the case, which is extremely convenient in making connections with circuits.

No fuses are placed in the converter. The primary fuses are placed within a primary switch and fuse box, which is always furnished with all transformers, and is to be placed in any convenient position where the fuses can be readily renewed. The secondary fuses are to be placed within the building to be lighted, or near the point where the secondary wires enter. The fuses being thus disposed, it follows that the transformer case need never be opened after the converter is installed.

This transformer is provided with the oil insulation already referred to in previous chapters. Glancing again at the illustration, Fig. 34, it will

Fig. 34.

at once be noted that the case is constructed in the form of a kind of cup or pocket, the only entrance being by means of a lid, or cap piece, at the top of the case. Within this complete receptacle the transformer is fixed, and when the apparatus is installed the case is filled with oil supplied for the purpose, thus completely surrounding and permeating the transformer with a very perfect self-renewing insulating medium. It is unnecessary to discuss the merits and advantages of oil insulation here, as they are fully considered in a previous chapter. The amount of oil required with each transformer varies from two quarts for a 600 watt converter, to nine quarts in the 7500 watt size.

Another safeguard against loss of insulation is added by the grounding of the core. The laminated iron is fully insulated from the case, and is connected thoroughly to earth by means of a ground wire. Thus a lightning discharge, entering the converter, would strike through the insulation from coil to lamination, and pass down the ground wire to earth, the puncture being immediately re-insulated by the oil. This matter was thoroughly discussed in Chapter VI.

These transformers are wound to stand a test of 5000 volts without oil insulation.

The case of the Thomson-Houston transformer is provided with hooks to permit of its being

Fig. 35.

readily installed upon cross-arms or transformer timbers on houses.

The merits claimed for this converter are high efficiency, perfect insulation, safety, long life, close regulation, small leakage, convenience and simplicity.

Previous to the introduction of the transformer just described, which is known as *Type F*, the Thomson-Houston Electric Co. manufactured a transformer known as their Type E, an illustration of which is shown in Fig. 35. This transformer is widely used and generally popular.

The Westinghouse Converter is shown in accompanying illustration, Fig. 36, and is manufactured by the Westinghouse Electric Co., of Pittsburg, Penn. Like most of the other commercial transformers it is block-shaped.

At the upper end of the case, and attached to the front plate, is a fuse and connection box. This contains a removable porcelain block, carrying the primary fuses. The removal of this block breaks the primary circuit, leaving the converter dead while the fuses are being renewed. New fuses may thus be placed with perfect safety and without trouble.

The fuse box is opened by removing the front plate, which is held in place by three thumb-screws, as shown by the illustration. The front plate cannot be dropped, being fastened to the converter by a chain. The secondary connections are made at the lower end of the case, being remote and thoroughly isolated from the primary circuit.

These transformers are manufactured in sizes ranging from 250 to 6250 watts, and to give sec-

FIG. 36.

ondary potentials of 50 or 100 volts, as is required. The manufacturers claim for them high efficiency, good regulation, high insulation, safety and careful construction.

In describing the foregoing transformers we have endeavored to confine ourselves strictly to their more striking and conspicuous features. A general description has, in every case, been rendered unnecessary by the excellent cuts which we have been enabled, through the kindness of the various manufacturers, to present.

Although this may justly be considered as the concluding chapter of this little work, there are yet to follow a number of pages devoted to sundry miscellaneous, but distinctly relevant, matters which, while too important to omit, could scarcely be classified under any of the previous headings.

APPENDICES.

1.—HIGH VOLTAGES. 2.—THE "HEDGEHOG" TRANSFORMER. 3.—THE WELDING MACHINE. 4.—DIRECT CURRENT TRANSFORMERS. 5.—STATION TRANSFORMERS. 6.—CONSTANT CURRENT TRANSFORMERS. 7.—UNDERWRITERS' RULES.

1. HIGH VOLTAGES.

REFERENCE has already been made to the use of very high pressures obtained by means of step-up transformers, for distributing large amounts of power through very long distances with relatively small loss.

The economy that would result from such a system has long been understood, but the many serious difficulties which have stood in the way of its successful application have, until recently, with isolated exceptions, caused its use to be confined almost entirely to experiment. Among the earliest experiments with high voltages from step-up transformers (as *distinct* from static pressures), were those conducted in London, just previous to the building of the celebrated Deptford station. Their purpose was to establish the feasibility of, and the necessary methods to be pursued in a system such as that which subsequently resulted.

In the frontispiece of this volume a photograph is re-produced, showing a discharge from the 10,000 volt circuit of a step-up transformer.

This experiment was to establish the distance that this pressure would force current through an ordinarily dry atmosphere.

The discharge depicted was between two copper points, separated from one another by five-eighths of an inch. Across this open space the enormous pressure caused the electric fluid to leap without difficulty.

Dry air being practically the best insulating medium known, the difficulty of properly insulating such high potentials will readily be appreciated. The flash at the right of the illustration is occasioned by the blowing of a fuse immediately upon the establishment of the circuit through the air. There were, at the time these experiments were first attempted, no thoroughly effective measuring instruments for computing what voltages really existed at such very high pressures.

The earliest method of definitely determining that 10,000 volts had actually been obtained was by illuminating 100 100-volt lamps in series. This required, of course, 10,000 volts between the two ends of the line of lamps, to bring them to full

130 TRANSFORMERS.

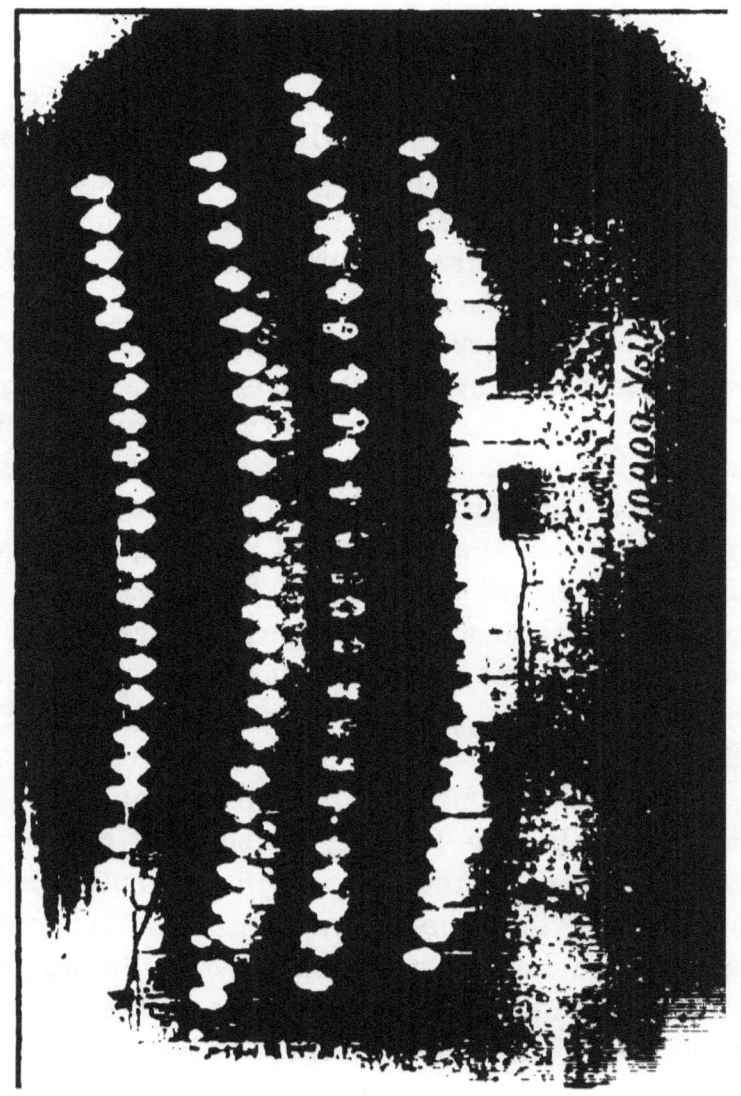

candle-power. Such an arrangement is shown in the accompanying illustration, Fig. 37.

Many recent experiments, conducted to ascertain the effects and results which could be obtained with very high potentials, combined, as a rule, with high frequency of alternations, have brought to our knowledge a large number of very beautiful and remarkable results.

Thus, with a very large induction coil or step-up transformer, having an enormous number of turns of very fine wire in the secondary, and a very rapid rate of alternation being used, the two ends of the secondary, though remote from one another, emit a more or less brilliant scintillating light, surrounding the ends of the wire. This resembles in general character and appearance that remarkable natural phenomenon, which at times appears at the ends of ships' masts and spars, known as St. Elmo's fire.

Incandescent lamps, having only a single thread of filament, become brightly illuminated on being attached to *one* of the wires, while on sufficiently intensifying these conditions a simple incandescent lamp of peculiar construction may be illuminated without being brought in contact with any wire or source of current whatever, and without being introduced into any magnetic field. This is accomplished by introducing the lamp between two metal plates, forming, perhaps, the sides of a small room, and each being attached to one terminal of the secondary already referred to.

An endless variety of experiments in this line have been made, chiefly by Mr. Nichola Tesla, to whom chiefly belongs the credit of investigation in this direction.

The results obtained are very interesting, and all tend to prove that the rules governing the action of electricity as we understand them, while fixed and certain, only apply within what are probably quite narrow limits. These experiments point to the future utilization of electricity for illuminating purposes in ways which are not as yet known to us, and which will probably be as greatly in advance of present systems as our existing methods are in advance of those of one hundred years ago.

Mr. Tesla has successfully reproduced, on a small scale, many of nature's marvelous electrical phenomena, and it is probable that yet more wonderful discoveries are still to come.*

2. THE "HEDGEHOG" TRANSFORMER.

The "Hedgehog" transformer is the invention of Mr. James Swinburne, the well-known English electrician, and was the outcome of his theory that

---

* For further information relative to these experiments we would refer the reader to the Proceedings of the American Institute of Electrical Engineers, 1891.

the iron and hysteresis losses were more serious in the transformer than the C_2R loss. He argued, therefore, that an open circuit transformer, with greatly reduced iron, would be more efficient, even though the copper would have to be greatly increased to obtain the same induction.

The iron loss is without doubt reduced in the construction adopted by Mr. Swinburne (See Chapter V), for not only is the sectional area of iron reduced, but being an open circuit transformer the outside iron is entirely wanting, thus reducing the total amount of iron to about one-third that used in a closed circuit transformer of similar capacity.

Thus far the theory is correct, for the "Hedgehog" transformer has shown, in careful tests, that its actual iron losses amount to considerably less than two per cent of full load. The change in the arrangement of iron, however, greatly increases the loss in the winding, more turns being needed to obtain the same results. So that the actual total efficiency of the "Hedgehog," according to Mr. Swinburne himself, proved to be about 87 per cent. To somewhat reduce the magnetic resistance incident to the use of an open iron circuit, the ends of the wire core are spread out as previously shown in Fig. 20, Chapter V, to equally distribute the magnetic lines through a broad path in the surrounding air, for it is generally accepted as true that the magnetic resistance of air is relatively low when the magnetic induction, or the magnetic current through a given area, is small.

The construction of the "Hedgehog" transformer is simple in the extreme. The iron wire of the core is built up around a brass, or gun metal back-bone, one end of which is spread out to form legs, and the other to hold a connection board. The windings being placed over the iron, the whole transformer is enclosed within an earthenware jar or case. Earthenware was selected for this purpose because, being an open circuit transformer, an iron case would at once become magnetized by the core, while a case of any metal would be subject to Foucault currents and would waste much energy.

3. THE ELECTRIC WELDING MACHINE.

An ingenious and valuable application of the transformer principle is the Electric Welding Machine. This is a transformer so constructed as to generate enormous current volume in the secondary, with very little pressure. The secondary coil consists, as a rule, of but one turn of heavy copper, generally a casting. One end of this is movable, being usually made in the form of a sliding carriage, moving upon the solid portion of one end of the secondary.

Each end of the secondary carries a clamp or vice. These clamps are placed directly opposite one another, one being fixed and the other moving with the sliding carriage. The two pieces of metal to be welded together being fixed in the clamps opposite and parallel, as required, the movable clamp is drawn forward toward the other, which brings the objects to be welded together into contact, completing the circuit of the secondary through them.

The current passing through the secondary turn being greatly in excess of the true carrying capacity of the articles to be welded together and, owing to their very small length between the clamps, their resistance not being sufficient to materially cut the current down, they are speedily heated to the melting point at the point of contact and highest resistance. The two ends are then forced together while molten, and the two objects are at once welded into one.

The welding machine is made in many styles, according to the purpose for which it is designed. Wire, chain, pipe, rod, axles, tires, projectiles, etc., are all welded successfully and with perfect ease. Cast, as well as wrought metals, may be welded, as also may unlike metals, such as copper and steel.

The machines are designed according to the work they are to do, the larger and more powerful ones containing many mechanical features of great interest. The closing of the weld is generally accomplished in the larger machines by means of hydraulic pressure, and water circulation is provided to keep the secondary and movable carriage thoroughly cool. The welding machine has a wide field of usefulness, and though but a recent invention has already become well known.

4. THE DIRECT CURRENT TRANSFORMER.

The statement has been made throughout this volume that electrical transformation could be accomplished only by means of alternating, pulsating or intermittent currents. This is, perhaps, not literally true, although practically so.

There is a system, however, whereby transformation is accomplished in connection with direct current, but only mechanically. Stated briefly, it consists simply of an electric motor, operating on a high potential circuit, and driving a dynamo which generates at any potential that may be determined upon. Thus, instead of energy being transformed directly from electrical power of one pressure to electrical power of another, it must be converted first into mechanical energy or motion, and this again to electrical power of the pressure required.

Machines have been designed whereby both transformations are accomplished in one piece of apparatus, both the motor (primary) and dynamo (secondary) windings being placed upon a single armature, the same fields serving both for dynamo and generator. The armature is, of course, provided with two commutators and sets of brushes, one for the motor and another for the dynamo winding, the two being as absolutely distinct as if carried upon separate armatures.

While necessarily somewhat less efficient than true electrical transformers of correct design, this electro-mechanical converter is, in certain cases, extremely useful and valuable, for, as has been previously stated, alternating current has not, generally speaking, been successfully applied to the distribution of power for mechanical purposes.

Thus, when it is desired to carry power for the operation of motors for long distances, as, for example, when a water power in a remote location is to be utilized to operate stationary motors, or street cars, then the energy may be transmitted as *direct current* of *high potential* through long distances, and over light wire, with small loss, being reduced to the required pressure by means of one or more motor-transformers situated at a convenient centre of distribution.

5. STATION TRANSFORMERS.

It is customary to have upon the switch-board of every alternating dynamo a small transformer, which is mounted upon a base or back, instead of being enclosed within a case, and these are known as station transformers.

FIG. 38.

These transformers are constructed with special care as regards regulation, etc., and have the same ratio of transformation as the transformers chiefly in use upon the line.

Volt-meters, to measure high potentials, are delicate and expensive, consequently the switch-board volt-meter is placed upon the secondary of the station transformer and indicates the potential of the secondary circuits. The volt-meter and the two or three lights which serve to illuminate the switch-board constitute the load carried by the station transformer, this load being, of course, practically constant.

An excellent illustration of a station transformer is shown in the accompanying cut, Fig. 38.

6. CONSTANT CURRENT TRANSFORMERS.

In addition to the transformers which have been mentioned, designed to operate upon constant currents, transformers have been successfully constructed to give a constant current and varying potential at the secondary, with the primary upon a constant potential circuit.

Such transformers are designed to operate arc lamps upon incandescent circuits. This result is obtained by especial design and construction, and further mention of it is purposely avoided, lest it should prove confusing and irrelevant.

7. UNDERWRITERS' RULES.

The following rules are an extract from the Rules and Regulations adopted by the New England Insurance Exchange and the Boston Fire Underwriters' Union, for electric lighting, as at present in force.

The rules in use throughout the various States of the Union vary somewhat from these in detail but are the same in general character. These regulations are of great importance and should be carefully noted and closely followed.

Only such rules are presented here as apply directly to the application of transformers. For the complete text of the regulations the reader is referred to the official publication.

SECONDARY GENERATORS OR CONVERTERS.

Converters must not be placed inside of any building. They may be placed on the outer walls when in plain sight and easy of access, but must be thoroughly insulated from them. If placed on wooden walls, or the woodwork of stone or brick buildings, the insulation must be fire-proof. When an underground service is used, the converter may be put in any convenient place that is dry and does not open into the interior of the building; this location must have the approval of the inspector before the current is turned on.

The converter should be enclosed in a metallic or non-combustible case.

If for any reason it becomes necessary that the primary wires leading to and from the converter should enter a building, they must be kept apart a distance of not less than twelve inches, and the same distance from all other conducting bodies. The insulation of the wire must be of the very best.

Safety fuses must be placed at the junction of all feeders and mains, and at the junction of mains and branches where necessary, also in both the primary and secondary wires of the converter, in such a manner as not to be affected by the heating of the coils. Secondary wires, after leaving the converter, will be subject to rules already given for services, inside wiring, etc.

Any provision for grounding the secondary circuit by means of "film cut-out" or other approved automatic device will be approved. A permanent ground will not be approved.

SECONDARY SYSTEMS.

In these systems where alternating currents of high electro-motive force are used on the main lines, and secondary currents of low electro-motive force are developed in local "converters" or "transformers," it is important that the entire primary circuit and the transformers should be excluded from any insured building, and be confined to the aerial line (the transformers being attached to the poles or the exterior of the buildings) or to underground conduits if such are used, or placed in fire-proof vaults or exterior buildings.

In those cases, however, where it may not be possible to exclude the transformers and entire primary from the building, the following precautions must be strictly observed:

The transformer must be constructed with or inclosed in a fireproof or incombustible case, and located at a point as near as possible to that at which the primary wires enter the building. Between these points the conductors must be heavily insulated with a coating of approved waterproof material and, in addition, must be so covered-in and protected that mechanical injury to them, or contact with them, shall be practically impossible.

These primary conductors, if within a building, must also be furnished with a double-pole switch, or separate switches on the ingoing and return wires and also with automatic double-pole cut-out where they enter the building or where they leave the main line, on the pole or in the conduit. The switches above referred to should, if possible, be inclosed in secure and fireproof boxes outside the building.

In the case of isolated plants using the secondary system, the transformers must be placed as near to the dynamos as possible, and all primary wires must be protected in the same manner as is indicated in the second paragraph above.

GLOSSARY.

ELECTRICAL TERMS NOT EXPLAINED IN THE PRECEDING CHAPTERS.

Ampere.—The unit of electric current. A current of just sufficient strength to deposit .005084 grains of copper per second.

Electro-Motive Force.—Generally expressed as E. M. F. (See Voltage.)

Kapp Lines.—A unit of lines of magnetic force.

Multiple.—Synonymous with Parallel. (See Chapters I and VI.)

Ohm.—The unit of electrical resistance. A resistance through which one volt can just force one ampere.

Parallel.—See Multiple.

Potential.—See Voltage.

Series.—Lamps or other pieces of apparatus are in series when the same and whole current passes through them all, one after another. (See Chapters I and VI.)

Voltage.—Synonyms: Potential, Electro-Motive Force, Pressure, etc. The number of volts pressure present at any given point in a circuit.

Watt.—The unit of electric power, or the volt-ampere. The volts multiplied by the amperes in use equal the watts.

INDEX.

A

Alternating current............................ 17, 19,	22
Analysis of iron.............................	82
Asbestos paper, use of.........................	89
Attraction, molecular...........................	52

B

Baking................................... 90, 108,	109
Banking in multiple............................	98
in series..............................	100
three wire............................	101
Binding bolts.................................	85
Blocks, terminal..............................	91
Box, transformer..............................	91
Burr on plates................................	83

C

Calorimeter.............................. 51,	53
Carbon.......................................	82
Capacity, transformer required..................	97
Case, transformer.............................	91
Coil, reactive............................ 39,	45
induction.............................	26
Faraday induction............... 64,	80
Ruhmkorff................... 64,	81
Coils, ends of.................................	89
Cobalt.......................................	40
Connections, secondary........................	108
Commutator, function of.......................	23
Constant current transformer...................	130
Construction, rules for.........................	94
Core, form of.................................	73
joining of................................	76

Core, open circuit.................................... 65, 81
    building of....................................... 84
    grounding of................................. 105, 123
    wire.. 80
Copper tape, use of................................... 87
Cost, influence of 67
Crossing of leads................................ 91, 94
Currents, magnetic.................................... 39

D

Danger from grounds.................................. 105
Deptford Station..................................... 128
Devices, safety....................................... 92
Dies for punching..................................... 82
Dick and Kennedy's early transformer.................. 67
Direct current.. 18
    transformers..................................... 136
Direction of current.................................. 15
Discharge, 10,000 volt............................... 128
Distribution, electrical.............................. 15
Dynamo, principle of.................................. 12

E

Eddy currents.................................... 12, 48
Efficiency of transformers............................ 62
    curve.. 61
Electron.. 14

F

Faraday coil..................................... 64, 80
Faraday's discoveries................................. 63
Ferranti converter............................... 72, 110
Ferranti's early transformer.......................... 66
Fiber, vulcanized..................................... 89
Field of force.. 12
Filings, iron... 40
Force, lines of.................................. 14, 40
Formulæ, Hopkinson's.................................. 55
Fort Wayne Electric Co............................... 118
Foucault currents................................ 12, 48
Friction, molecular................................... 53
Fuse, main circuit................................... 100
Fuses... 38
    location of................................. 106, 108
    replacing of................................. 92, 108

INDEX. 147

Fuses, rubber covered................................ 106, 108
    sub-circuit.. 100

G

General Electric Co.................................... 121
Gloves, rubber.. 108
Goulard and Gibbs, distribution....................... 66
Grounding of core................................. 105, 123

H

Hedgehog transformer....................... 72, 81, 132
High voltages.. 128
    measurement of.................................. 129
Humming of transformer.......................... 95, 107
Hysteresis.. 51

I

Improvement, direction of............................. 72
Induction, electrical........................ 10, 14, 17
    coil 26, 80
    self..................................... 32, 39, 46
    mutual.................................... 34, 46
Inertia, electrical............................... 10, 30
    magnetic.. 52
Intermittent current........................... 17, 19
Interrupter.. 25
Insulation....................................... 90, 94
Insulation in lamination............................. 84
    oil.................................... 92, 105, 123
    puncturing of............................. 92, 105
    of winding................................. 86, 89
    wood for.. 90
Iron, complete circuit of............................ 41
    analysis of..................................... 82
    English.. 82
    lamination of.................................. 50
    filings..................................... 40, 95
    permeability of................................ 40
    selection of................................... 82
    wrought, characteristics of.................... 43

J

Jablochkoff, distribution............................. 66
Jar, advantage of................................ 103, 108
Joints in windings................................... 88

K

Kapp lines	41
Kennedy's transformer	66

L

Lag	11
Lamp, incandescent	17
Lathe, winding	86
Lamination of iron	50
Law, Ohms	46
Leads, crossing of	91, 94
Lightning, effect of	104, 106
discharge, potential of	106
Lines of force	14, 40, 41
Kapp	41
Location of transformers	93, 103
Loss, C_2R	16
in converters	45
percentage of	54

M

Magnetic circuit, resistance of	66
currents	39
Magneto	93
-motive force	57
Mandril, winding	86
Mica, use of	89
Molecular attraction	52
friction	53
Multiple, banking in	98
arrangement	27
Mutual induction	34, 46

N

National Electrical Mfg. Co	115
transformer	115
Nickel	40

O

Ohm's law	46
Oil insulation	92, 105, 123

P

Paccinotti ring	87
Paper, asbestos	89

Paper, shellaced	89
Parallel arrangement	27
Phosphorus	82
Pine	90
Plates, punching of	82
Polarity, change of	52
Potential, best to use	98
difference of	10
Primary circuits, handling of	108
coil, character of 15,	17
Pulleys, tension	86
Pulsating currents 17, 19,	23
Punching of plates	82
Puncturing of insulation	105

R

Ratio of diameter to circumference	56
of windings	59
Reactive coil 39,	45
Regulation, self	34
Resistance of circuits	16
Ring-shaped core	78
Ribbon, silk, use of	87
Rubber tape, use of	87
Ruhmkorff coil 64,	81
Rules for construction	94
for installation	108
Underwriters' 108,	140

S

Saturation point	42
Safety devices	92
Secondary coil, character of 15,	17
Sectional area of wire	94
Self-induction 32, 39,	46
Series arrangement	27
Silicon	82
Slattery transformer	118
Stanley Electrical Manufacturing Co.	118
transformer	118
Stanley's early transformer	67
Steel, sheet	43
use of for cores	82
Step-up transformer 26, 107,	128
St. Elmo's fire	131
Station transformer	137

Swedish iron... 82
Swinburne's transformer........................... 72, 132
Symbols assumed.. 55

T

Tape, copper.... 87
      rubber... 87
Taping of coils.. 88
Teak, use of.. 90
Tension pulleys... 86
Terminal blocks... 91
Tesla's experiments.................................... 128
Testing of transformers................................ 93
Thomson-Houston transformer........................ 121
Three-wire arrangement............................... 101

U

Underwriters' rules............................... 108, 140

V

Vibration and hysteresis....................... 52, 103

W

Water, analogous to electricity..................... 16
Weight of transformer per useful watt............ 60
Welding machine....................................... 134
Westinghouse converter............................... 125
Winding, methods of........................... 76, 85
Windings, proportioning of......................... 60
      ratio of.. 59
Wire, coiling of.. 32
Wood for insulation................................... 90
Working point... 42

Z

Zipernowski and Derri's early transformer........ 70

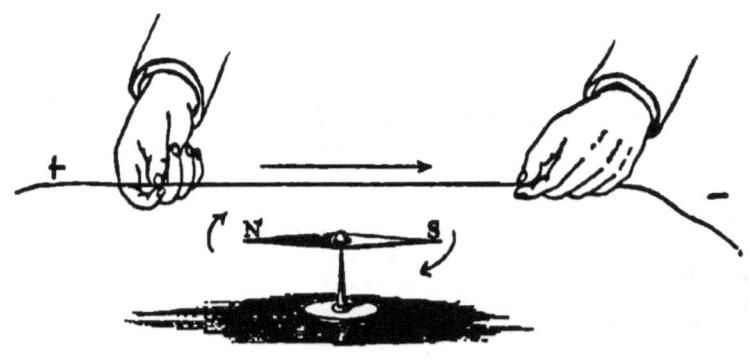

BUBIER'S
POPULAR ELECTRICIAN.

A Scientific Illustrated Monthly,

For the Amateur and Public at Large.

Containing descriptions of all the new inventions as fast as they are patented, also list of patents filed each month at the Patent Office in Washington, D. C. Interesting articles by popular writers on scientific subjects written in a way that the merest beginner in science can understand.

Price, Postpaid, 50 Cents a Year.

SAMPLE COPY, FIVE CENTS.

☞ Send for it.   You will be More than Pleased. ☜

Bubier Publishing Company, Lynn, Mass.

THE LEADING INSTRUCTION BOOKS UPON
ELECTRICITY.

"*Everybody's Hand-Book of Electricity,*" by Edward Trevert, with Glossary of Electrical Terms and Table for Incandescent Wiring. 120 Pages, 50 Illustrations. Giving a description of all the latest electrical inventions up to the present time. Over 25,000 sold already.

Price, Paper, 25 cents; Cloth, $1.00.

"*How to Make Electric Batteries at Home,*" by Edward Trevert, fully illustrated. The book gives just the information needed to make simple, yet practical, electrical batteries, by which you can run electric motors, incandescent lamps, bells, etc.

Price, 25 cents.

"*Experimental Electricity,*" by Edward Trevert. 176 Pages, 100 Illustrations. Giving complete working directions for making electric batteries, electric bells, induction coils, galvanometers, electric motors, dynamos, magnetos, etc.

Price, $1.00, Cloth Bound.

"*Dynamos and Electric Motors,*" *and All About Them,* by Edward Trevert. Fully Illustrated. This book gives complete directions for making Dynamos and Motors; also tells you all about them. Price, 50 cents, Cloth Bound.

"*Armature and Field-Magnet Winding,*" by Edward Trevert. Fully Illustrated with working drawings and fine engravings. All the information contained in this book is of a practical character, and the formulæ used is of the highest authority, by which one can wind a dynamo or motor for any output.

Price, $1.50, Cloth Bound.

"*How to Make a Dynamo,*" by Edward Trevert. Illustrated. Giving complete working directions and drawings for making a two incandescent light or small arc light machine. Price, 10 cents, Paper.

"*Electric Railway Engineering,*" by Edward Trevert. Illustrated. Embracing power house, dynamo, motor, and line construction up to date.

Price, $2.00, Cloth Bound.

"*Electricity and Its Recent Applications,*" by Edward Trevert. 350 pages. 250 Illustrations. 12mo. Cloth. The best and most complete book of the present time, and is particularly adapted for the use of students. Contains 20 chapters of the best and latest experiments, also complete working directions for building all kinds of electrical machinery. Price, $2.00, Cloth Bound.

"*A Practical Treatise on the Incandescent Lamp,*" by J. E. Randall, Electrician of the Incandescent Lamp Department of the Thompson-Houston Co. Illustrated. This is the only work that explains in a practical manner the manufacture of the Incandescent Lamp, and should be owned by every Electrician and student interested in the subject. Price, 50 cents, Cloth Bound.

"*A Practical Treatise on Electro-Plating,*" by Edward Trevert. Just the book for Amateurs. Fully Illustrated. Sent to any address on receipt of price.
Price, 50 cents, Cloth Bound.

"*Electric Motor Construction for Amateurs,*" by C. D. Parkhurst. Illustrated. Just the book for beginners or for anybody wishing to construct their own electrical apparatus. Giving complete directions and working drawings for making an electric motor for running sewing machines, small lathes, etc. Also gives directions and drawings for building an electric battery to furnish current for the motor. Price, $1.00, Cloth.

"*A Hand Book of Wiring Tables,*" for Arc, Incandescent Lighting and Motor Circuits, by A. E. Watson. This book gives a large number of Formulæ, Rules and Tables for Wiring, and is illustrated with numerous diagrams. It also contains a large amount of practical information upon the subject, up to date.
Price, 75 cents, Cloth.

"*How to Make and Use Induction Coils,*" by Edward Trevert. The only American book upon Induction Coils. It is fully illustrated, and every student of electricity should own one. Bound in a neat cloth binding.
Price, 50 cents, postpaid.

"*Questions and Answers About Electricity.*" A first book for students. Theory of Electricity and Magnetism. Edited by E. T. Bubier, 2d. *Authors*: T. O'Connor Sloane, A. M., E. M., PH. D.; Caryl D. Haskins, M. I. E. E.; A. E. Watson; Edward Trevert. Illustrated. *Contents*: Chapter I.— Theory of Electricity. II.— Theory of Magnetism. III.— Voltaic Batteries. IV.— Dynamos and Motors. V.— Electric Lamps. VI.— Miscellaneous Electrical Apparatus. VII.— Electrical Measurement.
Price, Cloth Bound, 50 cents.

All Books sent postpaid on receipt of price.
Send Money by P. O. order or registered letter at our risk.

Bubier Publishing Co.,
LYNN, MASS.

www.ingramcontent.com/pod-product-compliance
Lightning Source LLC
Chambersburg PA
CBHW030336170426
43202CB00010B/1146